Selkow

B

Progress in Computer Science and Applied Logic
Volume 8

Editor

John C. Cherniavsky, Georgetown University

Associate Editors

Robert Constable, Cornell University
Jean Gallier, University of Pennsylvania
Richard Platek, Cornell University
Richard Statman, Carnegie-Mellon University

Uwe Schöning

Logic for Computer Scientists

With 34 Illustrations

1989

Birkhäuser
Boston · Basel · Berlin

Professor Uwe Schöning
Universität Ulm
Abt. Theoretische Informatik
Oberer Eselsberg
D-7900 Ulm
Federal Republic of Germany

Library of Congress Cataloging-in-Publication Data
Schöning, Uwe, 1955-
    Logic for computer scientists / Uwe Schöning.
        p.   cm.—(Progress in computer science and applied logic :
    v. 8)
    Includes bibliographical references.
    ISBN 0-8176-3453-3 (alk. paper)
    1. Logic, Symbolic and mathematical.   2. Logic programming.
    I. Title.
    QA9.S363   1989
    511.3—dc20   89-17864

Logic for Computer Scientists was originally published in 1987
as *Logik für Informatiker* by Wissenschaftsverlag, Mannheim · Vienna · Zürich.

Printed on acid-free paper.

ISBN 0-8176-3453-3
ISBN 3-7643-3453-3

Camera-ready text provided by author using the LaTeX system of document preparation.
Printed and bound by R.R. Donnelley & Sons, Harrisonburg, Virginia.
Printed in the U.S.A.

9 8 7 6 5 4 3 2 1

Wolfram Schwabhäuser
(1931 – 1985)
in grateful memory

Preface

By the development of new fields and applications, such as Automated Theorem Proving and Logic Programming, Logic has obtained a new and important role in Computer Science. The traditional mathematical way of dealing with Logic is in some respect not tailored for Computer Science applications. This book emphasizes such Computer Science aspects in Logic. It arose from a series of lectures in 1986 and 1987 on Computer Science Logic at the EWH University in Koblenz, Germany. The goal of this lecture series was to give the undergraduate student an early and theoretically well-founded access to modern applications of Logic in Computer Science.

A minimal mathematical basis is required, such as an understanding of the set theoretic notation and knowledge about the basic mathematical proof techniques (like induction). More sophisticated mathematical knowledge is not a precondition to read this book. Acquaintance with some conventional programming language, like PASCAL, is assumed.

Several people helped in various ways in the preparation process of the original German version of this book: Johannes Köbler, Eveline and Rainer Schuler, and Hermann Engesser from B.I. Wissenschaftsverlag.

Regarding the English version, I want to express my deep gratitude to Prof. Ronald Book. Without him, this translated version of the book would not have been possible.

Koblenz, June 1989   U. Schöning

Contents

Introduction

Formal Logic investigates how assertions are combined and connected, how theorems formally can be deduced from certain axioms, and what kind of object a proof is. In Logic there is a consequent separation of syntactical notions (formulas, proofs) – these are essentially strings of symbols built up according to certain rules – and semantical notions (truth values, models) – these are "interpretations", assignments of "meanings" to the syntactical objects.

Because of its development from philosophy, the questions investigated in Logic were originally of a more fundamental character, like: What is truth? What theories are axiomatizable? What is a model of a certain axiom system?, and so on. In general, it can be said that traditional Logic is oriented to fundamental questions, whereas Computer Science is interested in what is programmable. This book provides some unification of both aspects.

Computer Science has utilized many subfields of Logic in areas such as program verification, semantics of programming languages, automated theorem proving, and logic programming. This book concentrates on those aspects of Logic which have applications in Computer Science, especially theorem proving and logic programming. From the very beginning, education in Computer Science supports the idea of strict separation between syntax and semantics (of programming languages). Also, recursive definitions, equations and programs are a familiar thing to a first year Computer Science student. This book is oriented in its style of presentation to this style.

In the first Chapter, propositional logic is introduced with emphasis on the resolution calculus and Horn formulas (which have their particular Computer Science applications in later sections). The second Chapter introduces the predicate logic. Again, Computer Science aspects are emphasized, like undecidability and semi-decidability of predicate logic, Herbrand's theory, and building upon this, the resolution calculus (and its refinements) for predicate logic is discussed. Most modern theorem proving programs are based on resolution refinements as discussed in Section 2.6.

The third Chapter leads to the special version of resolution (SLD-resolution) used in logic programming systems, as realized in the logic programming language PROLOG (= *Pro*gramming in *Log*ic). The idea of this book, though, is not to be a programmer's manual for PROLOG. Rather, the aim is to give the theoretical foundations for an understanding of logic programming in general.

Exercise 1: "What is the secret of your long life?" a centenarian was asked. "I strictly follow my diet: If I don't drink beer for dinner, then I always have fish. Any time I have both beer and fish for dinner, then I do without ice cream. If I have ice cream or don't have beer, then I never eat fish." The questioner found this answer rather confusing. Can you simplify it?

Find out which formal methods (diagrams, graphs, tables, etc.) you used to solve this Exercise. You have done your own first steps to develop a Formal Logic!

Chapter 1

PROPOSITIONAL LOGIC

1.1 Foundations

Propositional logic explores simple grammatical connections, like *and*, *or* and *not*, between the simplest "atomic sentences". Such atomic sentences are for example:

$$A = \text{"Paris is the capital of France"}$$
$$B = \text{"mice chase elephants"}$$

Such atomic components (of possibly more complex sentences) can be either *true* or *false*. (In our understanding of the world, A is true but B is false.) The subject of propositional logic is to declare formally how such "truth values" of the atomic components extend to a truth value of a more complex structure, such as

$$A \text{ and } B.$$

(For the above example, we know that A and B is false because B is already false.)

That is, we are interested in how the notion of a truth value extends from simple objects to more complex objects. In these investigations, we ignore what the underlying meaning of an atomic sentence is; our whole interest is concentrated on the truth value of the sentence.

3

For example, if

$$A = \text{``Charlie is getting sick''}$$
$$B = \text{``Charlie is consulting a doctor''}$$

then there is a big difference in colloquial language whether we say "A and B" or "B and A".

In the following definition we ignore such aspects occuring in natural language. All atomic sentences (now called atomic formulas) are thought of being enumerated as $A_1, A_2, A_3, \ldots$ ignoring the possible "meanings" of such formulas.

Definition (syntax of propositional logic)

An *atomic formula* has the form A_i where $i = 1, 2, 3, \ldots$. *Formulas* are defined by the following inductive process:

1. All atomic formulas are formulas.

2. For every formula F, $\neg F$ is a formula.

3. For all formulas F and G, also $(F \vee G)$ and $(F \wedge G)$ are formulas.

A formula of the form $\neg F$ is called *negation* of F. A formula of the form $(F \vee G)$ is called *disjunction* of F and G, and $(F \wedge G)$ is the *conjunction* of F and G. Any formula F which occurs in another formula G is called a *subformula* of G.

Example: $F = \neg((A_5 \wedge A_6) \vee \neg A_3)$ is a formula, and all subformulas of F are:

$$F, \ ((A_5 \wedge A_6) \vee A_3), \ (A_5 \wedge A_6), \ A_5, \ A_6, \ \neg A_3, \ A_3$$

We introduce the following abbreviations which allow a more succinct representation of formulas:

$$A, B, C, \ldots \quad \text{instead of} \quad A_1, A_2, A_3, \ldots$$
$$(F_1 \rightarrow F_2) \quad \text{instead of} \quad (\neg F_1 \vee F_2)$$
$$(F_1 \leftrightarrow F_2) \quad \text{instead of} \quad ((F_1 \wedge F_2) \vee (\neg F_1 \wedge \neg F_2))$$

$$(\bigvee_{i=1}^{n} F_i) \quad \text{instead of} \quad (\ldots((F_1 \vee F_2) \vee F_3) \vee \cdots \vee F_n)$$

$$(\bigwedge_{i=1}^{n} F_i) \quad \text{instead of} \quad (\ldots((F_1 \wedge F_2) \wedge F_3) \wedge \cdots \wedge F_n)$$

Here, $F_1, F_2, \ldots$ can be arbitrary formulas. In particular, that means that $(A \leftrightarrow E)$ is an abbreviation for the formula

$$((A \wedge E) \vee (\neg A \wedge \neg E))$$

which, again, is an abbreviation for

$$((A_1 \wedge A_5) \vee (\neg A_1 \wedge \neg A_5)).$$

Notice that formulas are nothing else but strings of symbols (i.e. syntactical objects). They do not have a "content" or "meaning" at the moment. Therefore, it would be incorrect (or premature) to read $\wedge$ as "and", and $\vee$ as "or". Better would be, say, "wedge" and "vee".

Formulas – and the components occuring in formulas – obtain an associated "meaning" by the following definition.

Definition (semantics of propositional logic)

The elements of the set $\{0, 1\}$ are called *truth values*. An *assignment* is a function $\mathcal{A} : \mathbf{D} \to \{0, 1\}$, where $\mathbf{D}$ is any subset of the atomic formulas. Given an assignment $\mathcal{A}$, we extend it to a function $\mathcal{A}' : \mathbf{E} \to \{0, 1\}$, where $\mathbf{E} \supseteq \mathbf{D}$ is the set of formulas that can be built up using only the atomic formulas from $\mathbf{D}$.

1. For every atomic formula $A_i \in \mathbf{D}$, $\mathcal{A}'(A_i) = \mathcal{A}(A_i)$.

2. $\mathcal{A}'((F \wedge G)) = \begin{cases} 1, & \text{if } \mathcal{A}'(F) = 1 \text{ and } \mathcal{A}'(G) = 1 \\ 0, & \text{otherwise} \end{cases}$

3. $\mathcal{A}'((F \vee G)) = \begin{cases} 1, & \text{if } \mathcal{A}'(F) = 1 \text{ or } \mathcal{A}'(G) = 1 \\ 0, & \text{otherwise} \end{cases}$

4. $\mathcal{A}'(\neg F) = \begin{cases} 1, & \text{if } \mathcal{A}'(F) = 0 \\ 0, & \text{otherwise} \end{cases}$

Since $\mathcal{A}'$ is an extension of $\mathcal{A}$ ($\mathcal{A}$ and $\mathcal{A}'$ agree on $\mathbf{D}$), from now on, we drop the distinction between $\mathcal{A}$ and $\mathcal{A}'$ and just write $\mathcal{A}$. (The reason for this temporary distinction was to be able to define $\mathcal{A}'$ formally.)

Example: Let $\mathcal{A}(A) = 1$, $\mathcal{A}(B) = 1$ and $\mathcal{A}(C) = 0$. Then we obtain:

$$
\begin{aligned}
\mathcal{A}(\neg((A \wedge B) \vee C)) \;&= \begin{cases} 1, & \text{if } \mathcal{A}(((A \wedge B) \vee C)) = 0 \\ 0, & \text{otherwise} \end{cases} \\[2mm]
&= \begin{cases} 0, & \text{if } \mathcal{A}(((A \wedge B) \vee C)) = 1 \\ 1, & \text{otherwise} \end{cases} \\[2mm]
&= \begin{cases} 0, & \text{if } \mathcal{A}((A \wedge B)) = 1 \text{ or } \mathcal{A}(C) = 1 \\ 1, & \text{otherwise} \end{cases} \\[2mm]
&= \begin{cases} 0, & \text{if } \mathcal{A}((A \wedge B)) = 1 \;\;(\text{ because } \mathcal{A}(C) = 0\ ) \\ 1, & \text{otherwise} \end{cases} \\[2mm]
&= \begin{cases} 0, & \text{if } \mathcal{A}(A) = 1 \text{ and } \mathcal{A}(B) = 1 \\ 1, & \text{otherwise} \end{cases} \\[2mm]
&= 0
\end{aligned}
$$

The (semantic) effect of the "operators" $\wedge, \vee, \neg$ can be described by the following tables.

$\mathcal{A}(F)$	$\mathcal{A}(G)$	$\mathcal{A}((F \wedge G))$
0	0	0
0	1	0
1	0	0
1	1	1

$\mathcal{A}(F)$	$\mathcal{A}(G)$	$\mathcal{A}((F \vee G))$
0	0	0
0	1	1
1	0	1
1	1	1

$\mathcal{A}(F)$	$\mathcal{A}(\neg F)$
0	1
1	0

Using these tables, it is easy to determine the truth value of a formula F, once an assignment of the variables occuring in F is given. As an example, we consider again the formula $F = \neg((A \wedge B) \vee C)$, and we represent the way F is built up by its subformulas as a tree:

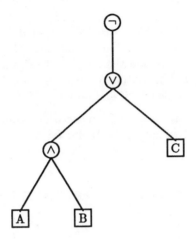

The truth value of F is obtainable by marking all leaves of this tree with the truth values given by the assignment $\mathcal{A}$, and then determining the values of the inner nodes according to the above tables. The mark at the root gives the truth value of F under the given assignment $\mathcal{A}$.

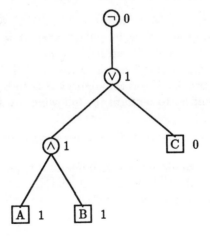

Exercise 2: Find a formula F containing the three atomic formulas A, B, and C with the following property: For every assignment $\mathcal{A} : \{A, B, C\} \to \{0, 1\}$, changing any of the values $\mathcal{A}(A)$, $\mathcal{A}(B)$, $\mathcal{A}(C)$ also changes $\mathcal{A}(F)$.

$(A \wedge B \wedge C) \vee (\neg A \wedge \neg B \wedge C) \vee (\neg A \wedge B \wedge \neg C) \vee (A \wedge \neg B \wedge \neg C)$

From the definition of $\mathcal{A}(F)$ it can be seen that the symbol "$\wedge$" is intended to model the spoken word "and", and similarly, "$\vee$" models "or", and "$\neg$" models "not". If we add the symbols "$\to$" and "$\leftrightarrow$" (which we introduced as syntactical abbreviations), then "$\to$" stands for "implies" or "if ... then", and "$\leftrightarrow$" stands for "if and only if".

To make the evaluation easier of formulas which contain the (abbreviation) symbols $\to$ or $\leftrightarrow$, we introduce tables for these symbols as above.

$\mathcal{A}(F)$	$\mathcal{A}(G)$	$\mathcal{A}((F \to G))$	$\mathcal{A}(F)$	$\mathcal{A}(G)$	$\mathcal{A}((F \leftrightarrow G))$
0	0	1	0	0	1
0	1	1	0	1	0
1	0	0	1	0	0
1	1	1	1	1	1

Remark (induction on the formula structure)

The definition of formulas is an inductive definition: First, the simplest formulas are defined (the atomic formulas), then it is shown how more complicated formulas can be built up from simpler ones. The definition of $\mathcal{A}(F)$ is also by induction on the formula structure. This induction principle can also be used in proofs: If some statement S is to be proved for every formula F, then it suffices to perform the following two steps.

1. *(Induction Base)* Show that S holds for every atomic formula A_i.

2. *(Induction Step)* Show under the (induction) hypothesis that S holds for (arbitrary, but fixed) formulas F and G, it follows that S also holds for $\neg F$, $(F \wedge G)$, and $(F \vee G)$.

Definition (suitable assignment, model, satisfiable, valid)

Let F be a formula and let $\mathcal{A}$ be an assignment, i.e. a mapping from a subset of $\{A_1, A_2, \ldots\}$ to $\{0, 1\}$. If $\mathcal{A}$ is defined for every atomic formula A_i occuring in F, then $\mathcal{A}$ is called *suitable* for F.

If $\mathcal{A}$ is suitable for F, and if $\mathcal{A}(F) = 1$, then we write $\mathcal{A} \models F$. In this case we say F *holds* under the assignment $\mathcal{A}$, or $\mathcal{A}$ is a *model* for F. Otherwise we write $\mathcal{A} \not\models F$, and say: under the assignment $\mathcal{A}$, F does not hold, or $\mathcal{A}$ is not a model for F.

A formula F is *satisfiable* if F has at least one model, otherwise F is called *unsatisfiable* or *contradictory*. Similary, a set **M** of formulas is satisfiable if there exists an assignment which is a model for every formula F in **M**. (Note that this implies that this assignment is suitable for every formula in **M**).

A formula F is called *valid* (or a *tautology*) if *every* suitable assignment for F is a model for F. In this case we write $\models F$, and otherwise $\not\models F$.

Theorem

A formula F is a tautology if and only if $\neg F$ is unsatisfiable.

Proof:

F is a tautology	**iff**	every suitable assignment for F is a model for F
	iff	every suitable assignment for F (hence also for $\neg F$) is not a model for $\neg F$
	iff	$\neg F$ does not have a model
	iff	$\neg F$ is unsatisfiable.

■

The step from F to $\neg F$ (or vice versa) can be visualized by the following "mirror principle":

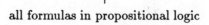

all formulas in propositional logic

valid formulas	satisfiable, but non-valid formulas	unsatis- fiable formulas
¬G	F ¬F	G

Application of the negation symbol means a reflection at the broken line. Hence a valid formula becomes an unsatisfiable formula (and vice versa), and a formula being satisfiable, but non-valid, again becomes a formula of this type.

Exercise 3: A formula G is called a *consequence* of a set of formulas $\{F_1, F_2, \ldots, F_k\}$ if for every assignment $\mathcal{A}$ which is suitable for each of $F_1, F_2, \ldots, F_k$ and G, it follows that whenever $\mathcal{A}$ is a model for $F_1, F_2, \ldots, F_k$, then it is also a model for G.

Show that the following assertions are equivalent:

1. G is a consequence of $\{F_1, F_2, \ldots, F_k\}$.

2. $((\bigwedge_{i=1}^{k} F_i) \rightarrow G)$ is a tautology.

3. $((\bigwedge_{i=1}^{k} F_i) \wedge \neg G)$ is unsatisfiable.

Exercise 4: What is wrong with the following argument?

"If I run the 100 meter race faster than 10.0 seconds, I will be admitted to the Olympic games. Since I am not running the 100 meter race faster than 10.0 seconds, I will not be admitted to the Olympic games."

The truth value of a formula obviously depends only on the truth assignments to the atomic formulas which occur in the formula. More formally, if two suitable assignments $\mathcal{A}$ and $\mathcal{A}'$ for F agree on all the atomic formulas which occur in F, then $\mathcal{A}(F) = \mathcal{A}'(F)$. (A formal proof of this fact would be by induction on the formula structure of F).

The conclusion we can draw is, for determining whether a given formula F is satisfiable or valid, it suffices to test finitely many different assignments for the atomic formulas occuring in F. If F contains the atomic formulas $A_1, \ldots, A_n$, then there are exactly 2^n different assignments (because there are 2^n different functions from $\{A_1, \ldots, A_n\}$ to $\{0, 1\}$). This test can be done systematically by *truth-tables*:

	A_1	A_2	$\cdots$	A_{n-1}	A_n	F
$\mathcal{A}_1$:	0	0		0	0	$\mathcal{A}_1(F)$
$\mathcal{A}_2$:	0	0		0	1	$\mathcal{A}_2(F)$
$\vdots$			$\ddots$			$\vdots$
$\mathcal{A}_{2^n}$:	1	1		1	1	$\mathcal{A}_{2^n}(F)$

It is clear now, that F is satisfiable if and only if the sequence of obtained truth values for F (the column below F) contains a 1, and F is valid if and only if the sequence consists only of 1's.

Example: Let $F = (\neg A \to (A \to B))$.

It is more convenient to have an extra column for every subformula occuring in F. Hence we obtain

A	B	$\neg A$	$(A \to B)$	F
0	0	1	1	1
0	1	1	1	1
1	0	0	0	1
1	1	0	1	1

The column below F consists only of 1's, therefore F is a tautology.

Remark: The truth-table method allows us to test formulas for satisfiability or for validity in a systematic, i.e. algorithmic way. But note that the expense of this algorithm is immense: For a formula containing n atomic formulas, 2^n rows of the truth-table have to be evaluated. For a formula with (only) 100 atomic formulas, the fastest existing computers would be busy for thousands of years to determine whether the formula is, say, satisfiable. (Find out how long 2^{100} microseconds are – supposing that one line of the truth-table can be constructed in 1 microsecond). This exponential behavior regarding the running time of potential algorithms for the satisfiability problem in propositional logic does not seem to be improvable (except for special cases, see Section 1.3). The satisfiability problem is "NP-complete". (This notion cannot be explained here, see any book on Complexity Theory).

Exercise 5: Show that a formula F of the form

$$F = (\bigwedge_{i=1}^{k} G_i)$$

is satisfiable if and only if the set of formulas $M = \{G_1, \cdots, G_k\}$ is satisfiable. Is this also true for formulas F of the form

$$F = (\bigvee_{i=1}^{k} G_i) \ ?$$

Exercise 6: How many different formulas F with the atomic formulas $A_1, \ldots, A_n$ and with different truth value sequences (columns below F) do there exist?

Exercise 7: Give an example of a 3-element set M so that M is not satisfiable, but every 2-element subset of M is satisfiable. Generalize your example to n-element sets.

Exercise 8: Is the following infinite set of formulas satisfiable?

$$M = \{A_1 \vee A_2, \neg A_2 \vee \neg A_3, A_3 \vee A_4, \neg A_4 \vee \neg A_5, \ldots\}$$

Exercise 9: Construct truth-tables for each of the following formulas.

$$((A \land B) \land (\neg B \lor C))$$

$$\neg(\neg A \lor \neg(\neg B \lor \neg A))$$

$$(A \leftrightarrow (B \leftrightarrow C))$$

Exercise 10: Prove or give a counter example:

(a) If $(F \to G)$ is valid and F is valid, then G is valid.

(b) If $(F \to G)$ is satisfiable and F is satisfiable, then G is satisfiable.

(c) If $(F \to G)$ is valid and F satisfiable, then G is satisfiable.

Exercise 11:

(a) Everybody having a musical ear is able to sing properly.

(b) Nobody is a real musician if he cannot electrify his audience.

(c) Nobody who does not have a musical ear can electrify his audience.

(d) Nobody, except a real musician, can compose a symphony.

Question: Which properties does a person have who has composed a symphony?

Formalize these assertions, and use truth-tables!

Exercise 12: Assume $(F \to G)$ is a tautology such that F and G do not share a common atomic formula. Show that either F is unsatisfiable or that G is a tautology (or both).

Show that the assumption about not sharing atomic formulas is necessary.

Exercise 13: (Craig's interpolation theorem)

Let $\models (F \rightarrow G)$ and let F and G have at least one atomic formula in common. Prove that there exists a formula H which is only built up from atomic formulas occuring in both F and G such that $\models (F \rightarrow H)$ and $\models (H \rightarrow G)$.

Hint: Use induction on the number of atomic formulas that occur in F, but not in G. Alternatively, construct a truth-table for H.

1.2   Equivalence and Normal Forms

From the way we assign truth values to formulas, we know that $(F \vee G)$ and $(G \vee F)$ "mean the same thing" – but syntactically the two formulas are different objects. We express this semantic equality or equivalence with the following definition.

Definition

Two formulas F and G are *(semantically) equivalent* if for every assignment $\mathcal{A}$ that is suitable for both F and G, $\mathcal{A}(F) = \mathcal{A}(G)$. Symbolically we denote this by $F \equiv G$.

Remark: Formulas containing different sets of atomic formulas can be equivalent (for example, tautologies).

Theorem (substitution theorem)

Let F and G be equivalent formulas. Let H be a formula with an occurence of F as subformula. Then H is equivalent to H' where H' is a formula obtained from H by substituting an occurence of subformula F by G.

Proof (by induction on the formula structure of H):

Induction Base: If H is an atomic formula with an occurence of F as subformula, then $H = F$. Therefore, $H' = G$ which is equivalent to H.

Induction Step: Let H be a non-atomic formula. In the case that the subformula F of H is H itself, the same argument as in the induction base applies. So suppose that $F \neq H$.

Case 1: H has the form $\neg H_1$.

The formula F is a subformula of H_1. Therefore, by induction hypothesis, H_1 is equivalent to H_1' where H_1' is obtained from H_1 by substituting F by G. Thus we have $H' = \neg H_1'$. By the (semantic) definition of "$\neg$" it follows that H and H' are equivalent.

Case 2: H has the form $(H_1 \vee H_2)$.

Then the occurence of F in H is either in H_1 or in H_2. We assume the former case in the following (the latter case is analogous). Then again, by induction hypothesis, H_1 is equivalent to H_1' where H_1' is obtained fron H_1 by substituting F by G. Using the semantic definition of "$\vee$" it is clear that $H \equiv (H_1' \vee H_2) = H'$.

Case 3: H has the form $(H_1 \wedge H_2)$.

This case is proved analogous to Case 2. ∎

Exercise 14: Let $F \equiv G$. Show: if F' and G' are obtained from F respectively G by substituting all occurences of $\vee$ by $\wedge$ (and vice versa) then $F' \equiv G'$.

Theorem

For all formulas F, G, and H, the following equivalences hold.

$$(F \wedge F) \equiv F$$
$$(F \vee F) \equiv F \qquad \text{(Idempotency)}$$

$$(F \wedge G) \equiv (G \wedge F)$$
$$(F \vee G) \equiv (G \vee F) \qquad \text{(Commutativity)}$$

$$((F \wedge G) \wedge H) \equiv (F \wedge (G \wedge H))$$
$$((F \vee G) \vee H) \equiv (F \vee (G \vee H)) \qquad \text{(Associativity)}$$

$$(F \wedge (F \vee G)) \equiv F$$
$$(F \vee (F \wedge G)) \equiv F \qquad \text{(Absorption)}$$

$$(F \wedge (G \vee H)) \equiv ((F \wedge G) \vee (F \wedge H))$$
$$(F \vee (G \wedge H)) \equiv ((F \vee G) \wedge (F \vee H)) \qquad \text{(Distributivity)}$$

$$\neg\neg F \equiv F \qquad \text{(Double Negation)}$$

$$\neg(F \wedge G) \quad \equiv \quad (\neg F \vee \neg G)$$
$$\neg(F \vee G) \quad \equiv \quad (\neg F \wedge \neg G) \qquad\qquad \text{(deMorgan's Laws)}$$

$$(F \vee G) \quad \equiv \quad F, \text{ if } F \text{ is a tautology}$$
$$(F \wedge G) \quad \equiv \quad G, \text{ if } F \text{ is a tautology} \qquad \text{(Tautology Laws)}$$

$$(F \vee G) \quad \equiv \quad G, \text{ if } F \text{ is unsatisfiable}$$
$$(F \wedge G) \quad \equiv \quad F, \text{ if } F \text{ is unsatisfiable} \quad \text{(Unsatisfiability Laws)}$$

Proof: All equivalences can be shown easily using the semantic definition of propositional logic. Also, we can verify them using truth tables. As an example we show this for the first absorption law.

$\mathcal{A}(F)$	$\mathcal{A}(G)$	$\mathcal{A}((F \vee G))$	$\mathcal{A}((F \wedge (F \vee G)))$
0	0	0	0
0	1	1	0
1	0	1	1
1	1	1	1

The first column and the fourth column coincide. Therefore, it follows

$$(F \wedge (F \vee G)) \quad \equiv \quad F .$$

■

Example: Using the above equivalences and the substitution theorem (ST) we can prove that

$$((A \vee (B \vee C)) \wedge (C \vee \neg A)) \equiv ((B \wedge \neg A) \vee C)$$

because we have

$$((A \vee (B \vee C)) \wedge (C \vee \neg A))$$
$$\equiv \quad (((A \vee B) \vee C) \wedge (C \vee \neg A)) \qquad \text{(Associativity and ST)}$$
$$\equiv \quad ((C \vee (A \vee B)) \wedge (C \vee \neg A)) \qquad \text{(Commutativity and ST)}$$
$$\equiv \quad (C \vee ((A \vee B) \wedge \neg A)) \qquad\qquad \text{(Distributivity)}$$
$$\equiv \quad (C \vee (\neg A \wedge (A \vee B))) \qquad \text{(Commutativity und ST)}$$
$$\equiv \quad (C \vee ((\neg A \wedge A) \vee (\neg A \wedge B))) \qquad \text{(Distributivity and ST)}$$
$$\equiv \quad (C \vee (\neg A \wedge B)) \qquad \text{(Unsatisfiability Law and ST)}$$
$$\equiv \quad (C \vee (B \wedge \neg A)) \qquad\qquad \text{(Commutativity and ST)}$$
$$\equiv \quad ((B \wedge \neg A) \vee C) \qquad\qquad\qquad \text{(Commutativity)}$$

Remark: The associativity law gives us the justification for a certain freedom in writing down formulas. For example, the notation

$$F = A \wedge B \wedge C \wedge D$$

refers to an arbitrary formula from the following list.

$$(((A \wedge B) \wedge C) \wedge D)$$
$$((A \wedge B) \wedge (C \wedge D))$$
$$((A \wedge (B \wedge C)) \wedge D)$$
$$(A \wedge ((B \wedge C) \wedge D))$$
$$(A \wedge (B \wedge (C \wedge D)))$$

Since all these formulas are equivalent to each other, from the semantic viewpoint it does not matter which of the formulas is referred to.

Exercise 15: Show that for every formula F there is an equivalent formula G which contains only the operators $\neg$ and $\rightarrow$. Show that there exists a formula having no equivalent one containing only the operators $\vee$, $\wedge$ and $\rightarrow$.

Exercise 16: Show (by induction) the following generalizations of deMorgan's law and of the distributivity laws.

$$\neg(\bigvee_{i=1}^{n} F_i) \equiv (\bigwedge_{i=1}^{n} \neg F_i)$$

$$\neg(\bigwedge_{i=1}^{n} F_i) \equiv (\bigvee_{i=1}^{n} \neg F_i)$$

$$((\bigvee_{i=1}^{m} F_i) \wedge (\bigvee_{j=1}^{n} G_j)) \equiv (\bigvee_{i=1}^{m}(\bigvee_{j=1}^{n}(F_i \wedge G_j)))$$

$$((\bigwedge_{i=1}^{m} F_i) \vee (\bigwedge_{j=1}^{n} G_j)) \equiv (\bigwedge_{i=1}^{m}(\bigwedge_{j=1}^{n}(F_i \vee G_j)))$$

Exercise 17: Using the equivalences of the theorem, show that the formula $((A \vee \neg(B \wedge A)) \wedge (C \vee (D \wedge C)))$ is equivalent to $(C \vee D)$.

$$C$$

Exercise 18: Formalize the following statements as formulas, and then show that they are equivalent.

(a) "If the child has temperature or has a bad cough and we reach the doctor, then we call him."

(b) "If the child has temperature, then we call the doctor provided we reach him, and, if we reach the doctor then we call him, if the child has a bad cough."

In the following we show that every formula – whether it is built up in a complicated way or not – can be transformed in an equivalent one which has a certain normal form. Even more, the above equivalences and the substitution theorem suffice for proving this.

Definition (normal forms)

A *literal* is an atomic formula or the negation of an atomic formula. (In the former case the literal is called *positive* and *negative* in the latter.)

A formula F is in *conjunctive normal form* (**CNF**) if it is a conjunction of disjunctions of literals, i.e.

$$F = (\bigwedge_{i=1}^{n}(\bigvee_{j=1}^{m_i} L_{i,j})) \,,$$

$$\text{where } L_{i,j} \in \{A_1, A_2, \ldots\} \cup \{\neg A_1, \neg A_2, \ldots\}$$

A formula F is in *disjunctive normal form* (**DNF**) if it is a disjunction of conjunctions of literals, i.e.

$$F = (\bigvee_{i=1}^{n}(\bigwedge_{j=1}^{m_i} L_{i,j})) \,,$$

$$\text{where } L_{i,j} \in \{A_1, A_2, \ldots\} \cup \{\neg A_1, \neg A_2, \ldots\}$$

Theorem

For every formula F there is an equivalent formula F_1 in **CNF** and an equivalent formula F_2 in **DNF**.

Proof (by induction on the formula structure of F):

Induction Base: If F is an atomic formula, then F is already in **CNF** as well as in **DNF**.

Induction Step: We distinguish between 3 cases.

Case 1: F has the form $F = \neg G$.
Then, by induction hypothesis, there are formulas G_1 in **CNF** and G_2 in **DNF** that are equivalent to G. Let

$$G_1 = (\bigwedge_{i=1}^{n} (\bigvee_{j=1}^{m_i} L_{i,j})) \ .$$

Application of deMorgan's law to $\neg G_1$ (in the generalized form, see Exercise 16) yields

$$F \equiv (\bigvee_{i=1}^{n} \neg (\bigvee_{j=1}^{m_i} L_{i,j})) \ ,$$

and finally,

$$F \equiv (\bigvee_{i=1}^{n} (\bigwedge_{j=1}^{m_i} \neg L_{i,j}))$$

which, by the double negation law, becomes

$$F \equiv (\bigvee_{i=1}^{n} (\bigwedge_{j=1}^{m_i} \overline{L_{i,j}}))$$

$$\text{where } \overline{L_{i,j}} = \begin{cases} A_k & \text{if } L_{i,j} = \neg A_k \\ \neg A_k & \text{if } L_{i,j} = A_k \ . \end{cases}$$

Therefore, we have obtained a formula in **DNF** equivalent to F. Analogously one can obtain from G_2 a formula in **CNF** equivalent to F.

Case 2: F has the form $F = (G \vee H)$.
By induction hypothesis, there are equivalent formulas to G and to H in **DNF** and in **CNF**. To obtain a formula in **DNF** equivalent to F, we simply combine the **DNF** formulas for G and H by $\vee$ (and then use associativity.)

To obtain a formula in **CNF** equivalent to F, we first choose formulas G_1 and H_1 in **CNF** equivalent to G and H. Let

$$G_1 = (\bigwedge_{i=1}^{n} G_i')$$

$$H_1 = (\bigwedge_{l=1}^{k} H_l')$$

where G_i' and H_l' are disjunctions of literals. Using the generalized distributivity law (Exercise 16), we obtain

$$F \equiv (\bigwedge_{i=1}^{n}(\bigwedge_{l=1}^{k}(G_i' \vee H_l')))$$

Using associativity, the get the form

$$F \equiv (\bigwedge_{i=1}^{n \cdot k} F_i')$$

where the F_i' are disjunctions of literals. Possible double occurences of literals within a disjunction, or double occurences of disjunctions can be eliminated using the idempotency laws. Also, if some of the disjunctions are tautologies (because they contain a literal together with its complement) then these disjunctions can be eliminated by the tautology law. This ultimately gives a formula in **CNF**.

Case 3: F has the form $F = (G \wedge H)$
This case is analogous to Case 2.   ■

The induction proof of the previous theorem hides a recursive algorithm to produce equivalent **DNF** and **CNF** formulas for a given formula. A more direct method to transform a formula into equivalent, say, **CNF** is the following.

Given: a formula F.

1. Substitute in F every occurence of a subformula of the form

$$\neg\neg G \quad \text{by} \quad G\,,$$
$$\neg(G \wedge H) \quad \text{by} \quad (\neg G \vee \neg H)\,,$$
$$\neg(G \vee H) \quad \text{by} \quad (\neg G \wedge \neg H)\,,$$

until no such subformulas occur.

2. Substitute in F each occurence of a subformula of the form

$$(F \vee (G \wedge H)) \quad \text{by} \quad ((F \vee G) \wedge (F \vee H)) \,,$$
$$((F \wedge G) \vee H) \quad \text{by} \quad ((F \vee H) \wedge (G \vee H)) \,,$$

until no such subformulas occur.

The resulting formula is in **CNF** (it still might contain superfluous, but permissible occurences of tautologies).

If a truth-table of a formula F is given or has been constructed, then there is another method to produce an equivalent formula in **DNF** or **CNF**.

To obtain an equivalent formula in **DNF** proceed as follows. Every line of the truth-table with the truth value 1 gives rise to a conjunction. The literals occuring in this conjunction are determined as follows: If for the assignment $\mathcal{A}$ that corresponds to this line we have $\mathcal{A}(A_i) = 1$ then A_i is inserted as literal, otherwise $\neg A_i$.

To obtain a formula in **CNF** equivalent to the given formula F with its truth-table, one has to interchange the roles of 0 and 1, and of disjunction and conjunction in the above instruction.

Example: A formula F is given with the following truth-table.

A	B	C	F
0	0	0	1
0	0	1	0
0	1	0	0
0	1	1	0
1	0	0	1
1	0	1	1
1	1	0	0
1	1	1	0

Then we obtain immediately an equivalent formula in **DNF**

$$(\neg A \wedge \neg B \wedge \neg C) \vee (A \wedge \neg B \wedge \neg C) \vee (A \wedge \neg B \wedge C) \,,$$

and also a formula in **CNF**

$$(A \vee B \vee \neg C) \wedge (A \vee \neg B \vee C) \wedge$$
$$(A \vee \neg B \vee \neg C) \wedge (\neg A \vee \neg B \vee C) \wedge (\neg A \vee \neg B \vee \neg C) \,.$$

Exercise 19: Given is the following formula

$$((\neg A \rightarrow B) \vee ((A \wedge \neg C) \leftrightarrow B)) \,.$$

Using any of the above methods, construct an equivalent formula in **DNF** and an equivalent one in **CNF**.

Observe that the formulas in **DNF** or **CNF** that are produced by the above methods are not necessarily the shortest possible ones. This problem, namely producing equivalent formulas in **DNF** or **CNF** that are as short as possible is interesting in digital circuit design. The shorter the formula, the fewer gates are needed for the circuit which realizes this formula. These issues are not the theme of this presentation.

Observe also that all the algorithms presented for producing **DNF** or **CNF** might produce an exponential "blow up" in the formula size. This blow up is caused by the applications of the distributive law. Each application roughly doubles the formula size. A formula with a short **DNF** presentation in general has a long **CNF** presentation and vice versa.

Exercise 20: Show that for every formula F there exists a formula G in **CNF** which can be constructed efficiently from F and has at most 3 literals per conjunction such that F is satisfiable if and only if G is satisfiable. (Note: it is not equivalence between F and G that is claimed here.) Further, the size of G is linear in the size of F.

Hint: The atomic formulas of G consist of those of F plus additional atomic formulas. These additional atomic formulas correspond to the inner nodes of the "structure tree" of F.

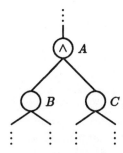

In this situation, the formula G would contain a subformula (transformed into **CNF**) of the form

$$\cdots \wedge (A \leftrightarrow (B \wedge C)) \wedge \cdots$$

The reader is invited to complete the details.

1.3 Horn Formulas

An important special case of **CNF** formulas which often occurs in practical applications are the Horn formulas (named after the logician Alfred Horn.)

Definition (Horn formula)

A formula F in **CNF** is a *Horn formula* if every disjunction in F contains at most one positive literal.

Example:

$$F = (A \vee \neg B) \wedge (\neg C \vee \neg A \vee D) \wedge (\neg A \vee \neg B) \wedge D \wedge \neg E$$

$$G = (A \vee \neg B) \wedge (C \vee \neg A \vee D).$$

F is a Horn formula and G is not.

Horn formulas can be (equivalently) rewritten in a more intuitive way, namely as implications. (We call this the procedural reading of Horn formulas.) In the above example, F can be rewritten as

$$F \equiv (B \rightarrow A) \wedge (C \wedge A \rightarrow D) \wedge (A \wedge B \rightarrow 0) \wedge (1 \rightarrow D) \wedge (E \rightarrow 0).$$

Here, 0 stands for an arbitrary unsatisfiable formula and 1 for an arbitrary tautology. It is easy to check that this equivalence really holds. The general rule is this: write the negative literals to the left of the implication sign (and a 1 if there is no negative literal), and write the positive literal (if any) at the right of the implication sign (and a 0 if there is no positive literal). Such an implication says whenever the premisses are satisfied, then the conclusion must be satisfied (and if the conclusion is 0, there is a contradiction). This informal argument will be made more formal in the following theorem.

A general theme of this book is the search for efficient algorithms which decide satisfiability (or validity) of formulas. Indeed, it is enough to have a test for unsatisfiability because a formula is valid if and only if its negation is unsatisfiable (cf. Exercise 3).

Using truth-tables, it is always possible to find out whether a formula is satisfiable or unsatisfiable. On the other hand, we have observed already that the expense of doing this is enormous: an algorithm based on constructing the full truth-table of a formula necessarily runs in exponential time.

In contrast, for Horn formulas there exists an efficient test for satisfiability which works as follows.

Instance: a Horn formula F

1. Mark every occurence of an atomic formula A in F if there is a subformula of the form $(1 \rightarrow A)$ in F.

2. **while** there is a subformula G in F of the form $(A_1 \wedge \cdots \wedge A_n \rightarrow B)$ or of the form $(A_1 \wedge \cdots \wedge A_n \rightarrow 0)$, $n \geq 1$, where $A_1, \ldots, A_n$ are already marked (and B is not yet marked) **do**

    if G is of the first form

        then mark every occurence of B

        else output "unsatisfiable" and halt ;

3. Output "satisfiable" and halt. (The satisfying assignment is given by the marking: $\mathcal{A}(A_i) = 1$ if and only if A_i has a mark.)

Theorem

The above marking algorithm is correct (for Horn formulas as input), and stops always after at most n many applications of the **while** loop ($n =$ number of atomic formulas in F.)

Proof: It is clear that the algorithm cannot mark more atomic formulas than there exist. Therefore, the output "satisfiable" or "unsatisfiable" is reached after at most n applications of the **while** loop.

Regarding the correctness of the algorithm, we observe that any model $\mathcal{A}$ for the input formula F (if there is any) must satisfy $\mathcal{A}(A_i) = 1$, for all atomic formulas A_i that are marked during application of the algorithm. This is immediate for the marked atomic formulas in step 1 of the algorithm because a CNF formula F obtains the truth value 1 only if *every* disjunction in F gets the value 1. If such a disjunction, as in step 1, has the form $(1 \rightarrow A)$, then A necessarily has to receive the assignment 1. Therefore, in step 2, it is necessary to mark (i.e. to assign 1 to) an atomic formula B provided $(A_1 \wedge \cdots \wedge A_n \rightarrow B)$ occurs in F and $A_1, \ldots A_n$ are already marked. Also, the decision for "unsatisfiable" is correct in the case that $(A_1 \wedge \cdots \wedge A_n \rightarrow 0)$ occurs in F and $A_1, \ldots A_n$ are already marked.

If the marking process successfully ends and step 3 is reached, then the formula F is satisfiable and the marking provides a model for F. To see this, let G be an arbitrary disjunction in F. If G is an atomic formula, then $\mathcal{A}(G) = 1$ is already guaranteed by step 1 of the algorithm. If G has the form $(A_1 \wedge \cdots \wedge A_n \rightarrow B)$ (i.e., $G = (\neg A_1 \vee \cdots \vee \neg A_n \vee B)$), then either every A_i is marked by 1, and by step 2 of the algorithm, also B is marked, or for at least one of the A_i, $\mathcal{A}(A_i) = 0$. In both cases we get $\mathcal{A}(G) = 1$. If G has the form $(A_1 \wedge \cdots \wedge A_n \rightarrow 0)$ (i.e., $G = (\neg A_1 \vee \cdots \vee \neg A_n)$), then, by the assumption that step 3 was reached, for at least one of the A_i, $\mathcal{A}(A_i) = 0$. Therefore, also in this case, $\mathcal{A}(G) = 1$.   ∎

Observe that the proof shows that the model $\mathcal{A}$ obtained by the marking is actually the smallest model for the formula F. That is, for every model $\mathcal{A}'$ and all atomic formulas B occuring in F, $\mathcal{A}(B) \leq \mathcal{A}'(B)$. (Here, the order $0 < 1$ is assumed.)

Another consequence of the proof is that every Horn formula is satisfiable if it does not contain a subformula of the form $(A_1 \wedge \cdots \wedge A_n \rightarrow 0)$. Exactly these subformulas possibly cause the above algorithm to halt with the output "unsatisfiable". Further, a Horn formula is satisfiable if it does not contain a subformula of the form $(1 \rightarrow A)$. In this case the **while** loop in step 2 will not be entered, and the control immediately reaches step 3.

Exercise 21: Apply the above marking algorithm to the Horn formula

$$F = (\neg A \vee \neg B \vee \neg D) \wedge \neg E \wedge (\neg C \vee A) \wedge C \wedge B \wedge (\neg G \vee D) \wedge G.$$

(Notice that a truth-table for this formula would have $2^6 = 64$ lines.)

Exercise 22: Give an example of a formula which does not have an equivalent Horn formula. Why is this so?

Exercise 23: Suppose we have the apparatuses available to perform the following chemical reactions.

$$\begin{aligned} MgO + H_2 &\rightarrow Mg + H_2O \\ C + O_2 &\rightarrow CO_2 \\ H_2O + CO_2 &\rightarrow H_2CO_3 \end{aligned}$$

Further, our lab has the following basic materials available: MgO, H_2, O_2 and C. Prove (by an appropriate application of the Horn formula algorithm) that under these circumstances it is possible to produce H_2CO_3.

1.4   The Compactness Theorem

In this section an important theorem is proved. Perhaps, the reader will not realize its importance at this time. But in Chapter 2 this theorem will play an important role.

Recall that a *set* M of formulas is, by definition, satisfiable if there is an assignment $\mathcal{A}$ such that for every $F \in M$, $\mathcal{A}(F) = 1$. We call such an assignment a *model* for M.

Compactness Theorem

A set M of formulas is satisfiable if and only if every *finite* subset of M is satisfiable.

Proof: Every model for M is also a model for every subset of M, in particular, for every finite subset of M. Therefore the direction from right to left is immediate.

Conversely, suppose that every finite subset of M is satisfiable, i.e. has a model. Our task is to construct one uniform model for M from this variety of models. For every $n \geq 1$ let M_n be the set of formulas in M that contains only the atomic formulas $A_1, \ldots, A_n$. Although M_n might be an infinite set, it contains at most 2^{2^n} many formulas with different truth tables. (Note that there are exactly 2^{2^n} many different truth tables with the atomic formulas $A_1, \ldots, A_n$). Therefore, there is a collection of formulas $\{F_1, \ldots, F_k\} \subseteq M_n$, $k \leq 2^{2^n}$, such that for every $F \in M_n$, $F \equiv F_i$ for some $i \leq k$. Hence, every model for $\{F_1, \ldots, F_k\}$ is also a model for M_n. By assumption, $\{F_1, \ldots, F_k\}$ possesses a model because it is a finite subset of M. Call this model $\mathcal{A}_n$. We further note that $\mathcal{A}_n$ is also a model for $M_1, \ldots, M_{n-1}$ because $M_1 \subseteq \cdots \subseteq M_{n-1} \subseteq M_n$.

We construct the desired model $\mathcal{A}$ for M in stages, such that we start with $\mathcal{A} = \emptyset$ in stage 0 and we declare in stage n how $\mathcal{A}$ is defined on A_n. Furthermore, in the construction appears an index set I which is initially set to $\mathbb{N}$, the set of all natural numbers, and modified at each stage. We find it convenient to use in some places the set theoretic notion for function and write $(A_n, 1) \in \mathcal{A}$ instead of $\mathcal{A}(A_n) = 1$. The stage construction follows.

Stage 0: $\mathcal{A} := \emptyset$;
$\quad\quad\quad I := \mathbb{N}$;
Stage $n > 0$: **if** there are infinitely many indices $i \in I$ with
$\quad\quad\quad\quad \mathcal{A}_i(A_n) = 1$ **then**
$\quad\quad\quad\quad\quad$ **begin**
$\quad\quad\quad\quad\quad\quad \mathcal{A} := \mathcal{A} \cup \{(A_n, 1)\}$;
$\quad\quad\quad\quad\quad\quad I := I - \{i \mid \mathcal{A}_i(A_n) \neq 1\}$
$\quad\quad\quad\quad\quad$ **end**
$\quad\quad\quad\quad$ **else**
$\quad\quad\quad\quad\quad$ **begin**
$\quad\quad\quad\quad\quad\quad \mathcal{A} := \mathcal{A} \cup \{(A_n, 0)\}$;
$\quad\quad\quad\quad\quad\quad I := I - \{i \mid \mathcal{A}_i(A_n) \neq 0\}$
$\quad\quad\quad\quad$ **end.**

Since in each stage n the assignment $\mathcal{A}$ is extended by $(A_n, 0)$ or by $(A_n, 1)$, but not both, $\mathcal{A}$ is a well-defined function with domain $\{A_1, A_2, A_3, \ldots\}$ and range $\{0, 1\}$.

We claim that $\mathcal{A}$ is a model for M. Let F be an arbitrary formula in M. F contains only finitely many atomic formulas, say, $A_1, \ldots A_l$. Therefore, F is an element of $M_l \subseteq M_{l+1} \subseteq \cdots$ and each of the assignments $\mathcal{A}_l, \mathcal{A}_{l+1}, \ldots$ is a model for F. It can be seen that the above construction has the property that in each stage, I is "thinned out" because indices are canceled from I, but I will never become finite. Therefore, in stage l infinitely many indices remain in I, also such indices i with $i \geq l$. All these remaining assignments $\mathcal{A}_i$ agree with each other and with $\mathcal{A}$ on $\{A_1, \ldots, A_l\}$. Hence, $\mathcal{A}(F) = 1$.

■

Observe that the above proof is *non-constructive*. That is, the *existence* of the model $\mathcal{A}$ is shown, but the test in the **if** statement cannot be checked in a finite amount of time (cf. Section 2.3 about decidability questions.) Rather, it is a "mental construction": either the **if** condition or the **else** condition is satisfied, and the construction is supposed to proceed correspondingly, but we are not able to implement this process algorithmically.

Formulated in different terms, the compactness theorem states that a set of formulas M is *unsatisfiable* if and only if there exists a *finite* subset of M that is *unsatisfiable*. In this form the compactness theorem will be used in Chapter 2. To give an understanding of this application in Chapter 2, suppose the set M can be enumerated by an algorithmic process

$$M = \{F_1, F_2, F_3, \ldots\},$$

that is, there is an algorithm which, on input n, outputs F_n in finite time. To determine whether M is unsatisfiable, we generate successively $F_1, F_2, F_3, \ldots$ and test each time whether the finite set of formulas generated so far is unsatisfiable. If so, we know that M is unsatisfiable. On the other hand, there is no way to confirm satisfiability in a similar manner.

Exercise 24: Let M be an infinite set of formulas so that every finite subset of M is satisfiable. Suppose, no formula in M contains the atomic formula A_{723}. Therefore suppose, that none of the assignments $\mathcal{A}_n$ in the above construction is defined on A_{723}. Find the value of $\mathcal{A}(A_{723})$ given by the above construction.

Exercise 25: Prove that $M = \{F_1, F_2, F_3, \ldots\}$ is satisfiable if and only if for infinitely many n, $(\bigwedge_{i=1}^{n} F_i)$ is satisfiable.

Exercise 26: A set of formulas M_0 is an *axiom system* for a set of formulas M if

$$\{\mathcal{A}|\mathcal{A} \text{ is model for } M_0\} = \{\mathcal{A}|\mathcal{A} \text{ is model for } M\}.$$

M is called *finitely axiomatizable* if M has a finite axiom system. Suppose, $\{F_1, F_2, F_3, \ldots\}$ is an axiom system for a set M where for all $n \geq 1$,

$$\models (F_{n+1} \to F_n) \text{ and } \not\models (F_n \to F_{n+1}).$$

Show that M is not finitely axiomatizable.

Exercise 27: Let L be an arbitrary infinite set of natural numbers, presented in binary notation (e.g., the set of prime numbers: $L = \{10, 11, 101, 111, 1011, \ldots\}$).) Prove there is an infinite sequence of different binary numbers $w_1, w_2, w_3, \ldots$ such that w_i is prefix of w_{i+1} and also prefix of some element of L.

1.5 Resolution

Resolution is a simple syntactic transformation applied to formulas. From two given formulas in a resolution step (provided resolution is applicable to the formulas), a third formula is generated. This new formula can then be used in further resolution steps, and so on.

A collection of such "mechanical" transformation rules we call a *calculus*. Mostly, a calculus (like resolution) has an easy algorithmic description, therefore a calculus is particularly qualified for computer implementation. In the case of resolution there is just one rule which is applied over and over again until a certain "goal formula" is obtained.

The definition of a calculus is sensible only if its *correctness* and its *completeness* can be established (both with respect to the particular task for which the calculus is designed). To be more precise in the case of the resolution calculus, the task is to prove *unsatisfiability* of a given formula. (Remember that many other questions about formulas can be reduced to unsatisfiability, cf. Exercise 3.)

In this case, correctness means that every formula for which the resolution calculus claims unsatisfiability indeed is unsatisfiable. Completeness means that for every unsatisfiable formula there is a way to prove this by means of the resolution calculus.

A general precondition for the application of resolution to a formula is that the formula (or set of formulas) is in **CNF**. That is, if necessary, the formula has to be transformed into an equivalent **CNF** formula (see also Exercise 20.) Let the formula F be

$$F = (L_{1,1} \vee \cdots \vee L_{1,n_1}) \wedge \cdots \wedge (L_{k,1} \vee \cdots \vee L_{k,n_k})$$

where the $L_{i,j}$ are literals, i.e. $L_{i,j} \in \{A_1, A_2, \cdots\} \cup \{\neg A_1, \neg A_2, \cdots\}$. For the presentation of resolution it is advantageous to represent formulas in **CNF** as sets of so-called *clauses* where a clause is a set of literals:

$$F = \{\{L_{1,1}, \ldots, L_{1,n_1}\}, \ldots, \{L_{k,1}, \ldots, L_{k,n_k}\}\}$$

In this example, $\{L_{1,1}, \ldots, L_{1,n_1}\}$ is a clause. Hence a clause corresponds to a disjunction. A comma separating two literals within a clause can be thought of $\vee$, whereas a comma that separates two clauses corresponds to a $\wedge$.

The elements in a set do not have an order or priority and multiple occurences of an element "melt" together into a single element. Therefore, simplifications stemming from associativity, commutativity or idempotency are "automatically" provided by the set notation. The following equivalent **CNF** formulas all have the same set presentation, namely $\{\{A_3\}, \{A_1, \neg A_2\}\}$:

$$((A_1 \vee \neg A_2) \wedge (A_3 \wedge A_3))$$
$$(A_3 \wedge (\neg A_2 \vee A_1))$$
$$(A_3 \wedge ((\neg A_2 \vee \neg A_2) \vee A_1))$$
etc.

To keep notation simple, in the following we use the same letter F to represent a **CNF** formula, and also its corresponding clause representation. Of course, the relationship between clause sets and formulas is not one to one, as the above example shows. Furthermore, we apply notions like equivalence and satisfiability also to clause sets.

Definition (resolvent)

Let C_1, C_2 and R be clauses. Then R is called a *resolvent* of C_1 and C_2 if there is a literal $L \in C_1$ such that $\overline{L} \in C_2$ and R has the form

$$R = (C_1 - \{L\}) \cup (C_2 - \{\overline{L}\}).$$

Here, $\overline{L}$ is defined as

$$\overline{L} = \left\{ \begin{array}{ll} \neg A_i & \text{if } L = A_i \text{ ,} \\ A_i & \text{if } L = \neg A_i \text{ .} \end{array} \right.$$

Graphically we denote this situation by the following diagram.

The above definition also includes the case that R is the empty set (if $C_1 = \{L\}$ and $C_2 = \{\overline{L}\}$ for some literal L.) This *empty clause* is denoted by the special symbol $\square$. By definition, the empty clause $\square$ is unsatisfiable. Therefore, a clause set which contains $\square$ as an element is unsatisfiable.

The following are some examples for resolutions.

$$\{A_3, \neg A_4, A_1\} \qquad \{A_4, \neg A_1\}$$
$$\{A_3, A_1, \neg A_1\}$$

$$\{A_3, \neg A_4, A_1\} \qquad \{A_4, \neg A_1\}$$
$$\{A_3, \neg A_4, A_4\}$$

$$\{A_{17}\} \qquad \{\neg A_{17}\}$$
$$\square$$

Exercise 28: Give the entire list of resolvents which can be generated from the set of clauses

$$\{\{A, E, \neg B\}, \{\neg A, B, C\}, \{\neg A, \neg D, \neg E\}, \{A, D\}\}.$$

Exercise 29: If R is a resolvent of two Horn clauses, justify that R is a Horn clause, too.

Resolution Lemma

Let F be a **CNF** formula, represented as set of clauses. Let R be a resolvent of two clauses C_1 and C_2 in F. Then, F and $F \cup \{R\}$ are equivalent.

Proof: Let $\mathcal{A}$ be an assignment that is suitable for F (and also for $F \cup \{R\}$). If $\mathcal{A} \models F \cup \{R\}$ then immediately, $\mathcal{A} \models F$. Conversely, suppose $\mathcal{A} \models F$, that is, for all clauses $C \in F$, $\mathcal{A} \models C$. Assume the resolvent R has the form $R = (C_1 - \{L\}) \cup (C_2 - \{\overline{L}\})$ where $C_1, C_2 \in F$ and $L \in C_1, \overline{L} \in C_2$.

Case 1: $\mathcal{A} \models L$.

Then, by $\mathcal{A} \models C_2$ and $\mathcal{A} \not\models \overline{L}$, it follows $\mathcal{A} \models (C_2 - \{\overline{L}\})$, and therefore $\mathcal{A} \models R$.

Case 2: $\mathcal{A} \not\models L$.

Then, by $\mathcal{A} \models C_1$, it follows $\mathcal{A} \models (C_1 - \{L\})$, and therefore $\mathcal{A} \models R$.   ∎

Definition

Let F be a set of clauses. Then $Res(F)$ is defined as

$$Res(F) = F \cup \{R \mid R \text{ is a resolvent of two clauses in } F\}.$$

Furthermore, define

$$\begin{aligned} Res^0(F) &= F \\ Res^{n+1}(F) &= Res(Res^n(F)) \text{ for } n \geq 0 . \end{aligned}$$

and finally, let

$$Res^*(F) = \bigcup_{n \geq 0} Res^n(F) .$$

Exercise 30: For the following set of clauses,

$$F = \{\{A, \neg B, C\}, \{B, C\}, \{\neg A, C\}, \{B, \neg C\}, \{\neg C\}\}$$

determine $Res^n(F)$ for $n = 0, 1, 2..$

Exercise 31: Prove that for every finite clause set F there is a $k \geq 0$ such that
$$Res^k(F) = Res^{k+1}(F) = \ldots = Res^*(F).$$
Estimate k (in terms of, e.g., the number of clauses, the maximum size of a clause, and and the number of different atomic formulas in F).

Exercise 32: Let F be a set consisting of n clauses that contains the atomic formulas $A_1, A_2, \ldots, A_n$. What is the maximum size of $Res^*(F)$?

Now we proceed to the proof of correctness and completeness of the resolution calculus (with respect to unsatisfiability). In this context, resolution is called *refutation complete.*

Resolution Theorem (of propositional logic)

A clause set F is unsatisfiable if and only if $\square \in Res^*(F)$.

Proof: Using the compactness theorem, we may assume that F is finite, otherwise we pick an unsatisfiable finite subset of F.

(Correctness) We need to show that $\square \in Res^*(F)$ implies that F is unsatisfiable. From the Resolution Lemma, we obtain
$$F \equiv Res^1(F) \equiv Res^2(F) \equiv \cdots \equiv Res^n(F) \equiv \cdots$$
Since $\square$ is contained in $Res^*(F)$, it is contained in $Res^n(F)$ for some n. The empty clause $\square$ can only be obtained from two clauses of the form $\{L\}$ and $\{\overline{L}\}$. Therefore, $\{L\}, \{\overline{L}\} \in Res^n(F)$. Obviously there is no assignment which can make all clauses in $Res^n(F)$ true, therefore, $Res^n(F)$ is unsatisfiable, and by the above equivalence, F is unsatisfiable.

(Completeness) Suppose that F is unsatisfiable. We show $\square \in Res^*(F)$ by induction on the number n of different atomic formulas in F.

Induction Base: If $n = 0$, then it must be that $F = \{\square\}$, and therefore, $\square \in Res^*(F)$.

Induction Step: Let n be arbitrary, but fixed. Suppose that for every unsatisfiable set of clauses G containing at most the atomic formulas $A_1, \ldots, A_n$, $\square \in Res^*(G)$. Let F be a clause set with the atomic formulas $A_1, \ldots, A_n$, A_{n+1}. Without loss of generality we may assume that no clause contains both A_{n+1} and $\neg A_{n+1}$ (Why?). From F we obtain two new clause sets F_0 and F_1 as follows. F_0 results from F by canceling every occurence of the positive literal A_{n+1} within a clause, and for every occurence of the negative literal $\neg A_{n+1}$ within a clause, the entire clause is canceled. Analogously F_1 is defined where the roles of A_{n+1} and $\neg A_{n+1}$ are interchanged.

Note that F_0 (F_1) essentially results from F by fixing the assignment of A_{n+1} to 0 (to 1, resp.) Therefore, both F_0 and F_1 are unsatisfiable. Assume to the contrary that F_0 has a satisfying assignment $\mathcal{A} : \{A_1, \ldots, A_n\} \to \{0, 1\}$. Then, $\mathcal{A}'$ is a satisfying assignment for F where

$$\mathcal{A}'(B) = \begin{cases} \mathcal{A}(B) & \text{if } B \in \{A_1, \ldots, A_n\} \\ 0 & \text{if } B = A_{n+1}. \end{cases}$$

This contradicts the unsatisfiability of F. Similarly it can be shown that F_1 is unsatisfiable.

Therefore, by induction hypothesis, $\square \in Res^*(F_0)$ and $\square \in Res^*(F_1)$. This means there is a sequence of clauses $C_1, C_2, \ldots, C_m$ such that

$C_m = \square$,
and for $i = 1, \ldots, m$, $C_i \in F_0$ or C_i is a resolvent of two clauses C_a, C_b with $a, b < i$.

An analogous sequence $C_1', C_2', \ldots, C_t'$ exists for F_1. Some of the clauses C_i were obtained from F by canceling the literal A_{n+1}. By restoring the original clauses $C_i \cup \{A_{n+1}\}$, and carrying A_{n+1} along in the resolution steps, we obtain from $C_1, C_2, \ldots, C_m$ a new "proof sequence" for F which witnesses that $\{A_{n+1}\} \in Res^*(F)$. Similarly, restoring $\neg A_{n+1}$ in the sequence $C_1', C_2', \ldots, C_t'$ shows that $\{\neg A_{n+1}\} \in Res^*(F)$.

By a further resolution step,

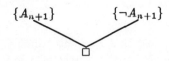

the empty clause can be derived, and therefore $\square \in Res^*(F)$.   ■

From the resolution theorem the following algorithm can be derived that decides satisfiability for a given input formula in **CNF** (or clause set) F (cf. Exercise 31).

Instance: a formula F in **CNF**

    1. Form a clause set from F (and continue to call it F);

    2. **repeat**
        $G := F$;
        $F := Res(F)$;
    until $(\square \in F)$ **or** $(F = G)$;

    3. **if** $\square \in F$ **then** "F is unsatisfiable"
    else "F is satisfiable";

In some cases this algorithm can come up with a decision quite fast, but there do exist examples for unsatisfiable formulas where exponentially many resolvents have to be generated before the **until** condition is satisfied (cf. Urquhart in the references).

In the following we want to distinguish between the clauses which are generated by the algorithm and those clauses thereof which are really relevant to derive the empty clause. (This might be significantly less clauses.) Implicitly, we used the following definition already in the proof of the resolution theorem.

Definition

A *derivation* (or *proof*) of the empty clause from a clause set F is a sequence $C_1, C_2, \ldots, C_m$ of clauses such that

C_m is the empty clause, and for every $i = 1, \ldots, m$, C_i either is a clause in F or a resolvent of two clauses C_a, C_b with $a, b < i$.

Reformulating the resolution theorem, it should be clear that a clause set F is unsatisfiable if and only if a derivation of the empty clause from F exists. To *prove* that a clause set F is unsatisfiable it is therefore enough to present a sequence of clauses according to the above definition. It is not necessary to write down *all* the clauses in $Res^*(F)$.

Example: Let $F = \{\{A, B, \neg C\}, \{\neg A\}, \{A, B, C\}, \{A, \neg B\}\}$. F is unsatisfiable. This fact is proved by the following derivation $C_1, \ldots, C_7$ where

$$
\begin{array}{rcll}
C_1 & = & \{A, B, \neg C\} & \text{(clause in } F) \\
C_2 & = & \{A, B, C\} & \text{(clause in } F) \\
C_3 & = & \{A, B\} & \text{(resolvent of } C_1, C_2) \\
C_4 & = & \{A, \neg B\} & \text{(clause in } F) \\
C_5 & = & \{A\} & \text{(resolvent of } C_3, C_4) \\
C_6 & = & \{\neg A\} & \text{(clause in } F) \\
C_7 & = & \square & \text{(resolvent of } C_5, C_6)
\end{array}
$$

This situation can be visualized by the *resolution graph*:

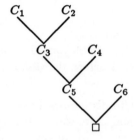

Such graphs need not necessarily be trees if the same clause is used in more than one resolution step.

Exercise 33: Using resolution, show that $A \wedge B \wedge C$ is a consequence of the clause set

$$F = \{\{\neg A, B\}, \{\neg B, C\}, \{A, \neg C\}, \{A, B, C\}\}.$$

Exercise 34: Using resolution, show that

$$F = (\neg B \wedge \neg C \wedge D) \vee (\neg B \wedge \neg D) \vee (C \wedge D) \vee B$$

is a tautology.

Exercise 35: Show that the following restriction of the resolution calculus is complete for the class of Horn formulas (but not for the general case): Derive a resolvent from two clauses C_1, C_2 only if one of these clauses is a *unit* clause, i.e. it consists of only one literal.

This resolution restriction has the property that the resolvents become shorter. Therefore, from the completeness of this restriction a similarly efficient algorithm for Horn formulas can be derived as the one presented in Section 1.3.

Hint: Show that the process of the marking algorithm for Horn formulas from Section 1.3 can be simulated in a certain way by appropriate applications of resolution steps with unit clauses.

Second Hint: This exercise will be solved in Section 2.6.

Exercise 36: Let F be a clause set with the atomic formulas $A_1, \ldots, A_n$ where each clause contains at most two literals (such clauses are called Krom clauses). How large can $Res^*(F)$ be at most? (From this exercise it follows that there is an efficient algorithm for determining satisfiability of Krom formulas.)

Exercise 37: Develop an efficient implementation of the resolution calculus which uses the following data structure: The example clauses

$$\{A, \neg B, C, D\}, \{A, B\}, \{\neg A, \neg B, \neg C\}, \{\neg B\}$$

are represented by the following *clause graph*,

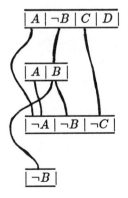

where an edge indicates a pair of complementary literals (and therefore the possibility of producing a resolvent.) Each edge can give the cause for a resolution step. In case a resolution step is performed, a new vertex representing the resolvent is generated. The edge connections to this new vertex can be read off from the parent vertices.

Furthermore, it is possible to cancel certain edges from the graph (and the neccessity to produce the corresponding resolvents) by certain locally checkable conditions. For example, both edges between the second and third clause can be canceled. Also, under certain conditions, vertices can be canceled from the graph, and need not be considered. For example, the first vertex can be canceled.

Exercise 38: Given is the following resolution graph where $C_1,\dots,C_7$ are *Horn* clauses.

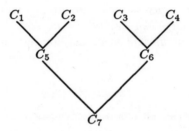

Show that this tree can be made linear, such that the clause C_7 can be obtained from C_1, C_2, C_3, C_4 in the following way

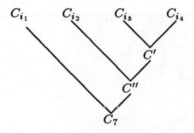

where $\{i_1, i_2, i_3, i_4\} = \{1, 2, 3, 4\}$ and C', C'' are certain suitably chosen Horn clauses.

Exercise 39: A clause is called *positive* (*negative*) if it contains only positive (negative) literals. Show that a clause set is satisfiable if it does not contain a positive clause. (The same holds if it does not contain a negative clause.)

Exercise 40: Show that the following restriction of resolution is complete: A resolvent of two clauses C_1, C_2 is only produced if one of the parent clauses is positive.

Hint: This exercise is solved in Section 2.6.

Exercise 41: Let F be an unsatisfiable clause set , and let G be a *minimally unsatisfiable* subset of F. (That means that G is unsatisfiable, but every proper subset of G is satisfiable.) Show that every derivation of the empty clause from F consists of at least $|G| - 1$ many resolution steps where $|G|$ denotes the number of clauses in G.

Remark: We have seen that in some special cases the resolution calculus leads to an efficient algorithm to determine (un)satisfiablity (cf. Exercises 35,36). But in the case of arbitrary clause sets, it is possible to exhibit unsatisfiable clause sets such that every derivation of the empty clause consists of exponentially many resolution steps (cf. Urquhart). That is, the expense of the resolution algorithm is comparable with the expense of the truth-table method. Because of the "NP-completeness" of the satisfiability problem, there does not seem to exist any significantly faster algorithm.

Another pecularity is worth mentioning: Both satisfiability and unsatisfiablity of a given formula F can be expressed by an *existential* statement. By definition, F is satisfiable if there *exists* a satisfying assignment for F. On the other hand, F is unsatisfiable if there *exists* a resolution derivation of the empty clause from F. As discussed above, there is a catch to this apparent symmetry. Writing down a resolution proof can be much more expensive than writing down a satisfying assignment. (This non-symmetry is closely related with the "NP=?co-NP" problem.)

Chapter 2

PREDICATE LOGIC

2.1 Foundations

Predicate logic can be understood as an extension of propositional logic. The additional new concepts include quantifiers, function symbols and predicate symbols. These new notions allow us to describe assertions which cannot be expressed with the available tools of propositional logic. For example, up to this point it was not possible to express that certain "objects" stand in certain relations, or that a property holds *for all* such objects, or that some object with a certain property *exists*. Here is a well known example from calculus:

> For all $\varepsilon > 0$ there exists some n_0, such that for all $n \geq n_0$, $abs(f(n) - a) < \varepsilon$.

The main concepts here are the verbal constructs *for all* and *exists*, as well as the use of functions (*abs*, f, $-$) and relations ($>$, $\geq$, $<$).

As in propositional logic, we start by formalizing the syntactic framework in which we want to discuss formulas in predicate logic. But first we need to define the syntax of the so-called *terms*, since terms occur as parts of formulas in predicate logic.

Definition (syntax of predicate logic)

A *variable* is of the form x_i where $i = 1, 2, 3, \ldots$. A *predicate symbol* is of

41

the form P_i^k and a *function symbol* of the form f_i^k where $i = 1, 2, 3, \ldots$ and
$k = 0, 1, 2, \ldots$. Here, i is the *distinguishability index* and k is called the
arity. In the case of arity 0, we drop the parentheses, and just write P_i^0 or
f_i^0. A function symbol of arity 0 will also be called a *constant*. Next, we
define *terms* by an inductive process as follows.

1. Each variable is a term.

2. If f is a function symbol with arity k, and if $t_1, \ldots, t_k$ are terms, then
   $f(t_1, \ldots, t_k)$ is a term.

Next, *formulas* (of predicate logic) are defined inductively as follows.

1. If P is a predicate symbol with arity k, and if $t_1, \ldots, t_k$ are terms,
   then $P(t_1, \ldots, t_k)$ is a formula.

2. For each formula F, $\neg F$ is a formula.

3. For all formulas F and G, $(F \wedge G)$ and $(F \vee G)$ are formulas.

4. If x is a variable and F is a formula, then $\exists x F$ and $\forall x F$ are formulas.

Atomic formulas are exactly those formulas built up according to rule 1. If
F is a formula, and F occurs as part of the the formula G, then F is called
a *subformula* of G.

All occurences of a variable in a formula are distinguished into *bound* and
free occurences. An occurence of the variable x in the formula F is bound
if x occurs within a subformula of F of the form $\exists x G$ or $\forall x G$. (Hence, the
same variable x can occur both free and bound in a formula F, see also
Exercise 42).

A formula without occurence of a free variable is called *closed*. The
symbols $\exists$ and $\forall$ are called *quantifiers* where $\exists$ is the *existential quantifier*
and $\forall$ is the *universal quantifier*. The *matrix* of a formula F, denoted
symbolically by F^*, is obtained by cancelling in F every occurence of a
quantifier and the variable that follows the quantifier.

Example: $F = (\exists x_1 P_5^2(x_1, f_2^1(x_2)) \vee \neg \forall x_2 P_4^2(x_2, f_7^2(f_4^0, f_5^1(x_3))))$ is a formula. All the subformulas of F are:

$$F$$
$$\exists x_1 P_5^2(x_1, f_2^1(x_2))$$
$$P_5^2(x_1, f_2^1(x_2))$$
$$\neg \forall x_2 P_4^2(x_2, f_7^2(f_4^0, f_5^1(x_3)))$$
$$\forall x_2 P_4^2(x_2, f_7^2(f_4^0, f_5^1(x_3)))$$
$$P_4^2(x_2, f_7^2(f_4^0, f_5^1(x_3)))$$

All the terms that occur in F are:

$$x_1$$
$$x_2$$
$$f_2^1(x_2)$$
$$f_7^2(f_4^0, f_5^1(x_3))$$
$$f_4^0$$
$$f_5^1(x_3)$$
$$x_3$$

All occurences of x_1 in F are bound. The first occurence of x_2 is free, all others are bound. Further, x_3 occurs free in F. Hence, the formula F is not closed. The term f_4^0 is an example for a constant. The matrix of F is the formula

$$F^* = (P_5^2(x_1, f_2^1(x_2)) \vee \neg P_4^2(x_2, f_7^2(f_4^0, f_5^1(x_3))))$$

Exercise 42: Let $Free(F)$ be the set of all variables that occur free in F. Define $Free(F)$ formally (by induction on the term and formula structure).

Again, we allow the same simplifying notations for formulas as in propositional logic. Additionally, we allow the following abbreviations.

u, v, w, x, y, z	always stand for variables.
a, b, c	always stand for constants.
f, g, h	stand for function symbols where the arity can always be inferred from the context.
P, Q, R	stand for predicate symbols where the arity can always be inferred from the context.

Exercise 43: List all subformulas and terms that occur in the formula

$$F = (Q(x) \lor (\exists x \forall y (P(f(x), z) \land Q(a)) \lor \forall x R(x, z, g(x)))))$$

Which subformulas are closed? Determine for each occurence of a variable if it is free or bound. What is the matrix of F?

To interprete formulas of predicate logic (i.e. to give them a semantics, i.e. a "meaning"), we need to associate functions to the function symbols and predicates to the predicate symbols (in both cases, we also have to fix some ground set on which the functions and predicates are defined). Furthermore, variables that occur free in a formula need to be interpreted as elements of the ground set. If this is done, the formula gets a "meaning", in this case, a truth value. This intuitive explanation will be made formal in the following definition.

Definition (semantics of predicate logic)

A *structure* is a pair $\mathcal{A} = (U_{\mathcal{A}}, I_{\mathcal{A}})$ where $U_{\mathcal{A}}$ is an arbitrary, non-empty set and is called the *ground set* or *universe*. Further, $I_{\mathcal{A}}$ is a mapping that maps

- each k-ary predicate symbol P to a k-ary predicate on $U_{\mathcal{A}}$ (if $I_{\mathcal{A}}$ is defined on P).

- each k-ary function symbol f to a k-ary function on $U_{\mathcal{A}}$ (if $I_{\mathcal{A}}$ is defined on f).

- each variable x to an element of $U_{\mathcal{A}}$ (if $I_{\mathcal{A}}$ is defined on x).

In other words, the domain of I_A is a subset of $\{P_i^k, f_i^k, x_i \mid i = 1, 2, 3, \ldots$ and $k = 0, 1, 2, \ldots\}$, and the range of I_A is a subset of all predicates, functions, and single elements of U_A. In the following, we abbreviate the notation and write P^A instead of $I_A(P)$, f^A instead of $I_A(f)$, and x^A instead of $I_A(x)$.

Let F be a formula and $\mathcal{A} = (U_A, I_A)$ be a structure. $\mathcal{A}$ is called *suitable* for F if I_A is defined for all predicate symbols, function symbols, and for all variables that occur free in F.

Example: $F = \forall x P(x, f(x)) \land Q(g(a, z))$ is a formula. Here, P is a binary and Q a unary predicate, f is unary, g a binary, and a a 0-ary function symbol. The variable z is free in F. An example for a structure $\mathcal{A} = (U_A, I_A)$ which is suitable for F is the following.

$$
\begin{aligned}
U_A &= \{0, 1, 2, 3, \ldots\} = \mathbb{N}, \\
I_A(P) &= P^A = \{(m, n) \mid m, n \in U_A \text{ and } m < n\}, \\
I_A(Q) &= Q^A := \{n \in U_A \mid n \text{ is prime }\} \\
I_A(f) &= f^A = \text{ the successor function on } U_A, \\
&\qquad \text{hence } f^A(n) = n + 1, \\
I_A(g) &= g^A = \text{ the addition function on } U_A, \\
&\qquad \text{hence } g^A(m, n) = m + n, \\
I_A(a) &= a^A = 2, \\
I_A(z) &= z^A = 3.
\end{aligned}
$$

In this structure F is obviously "true" (we will define this notion in a moment), because every natural number is smaller than its successor, and the sum of 2 and 3 is a prime number.

Of course, for this formula F one can also define suitable structures in which F is "false". That is, F is not a "valid" formula, i.e. F is not true in every suitable structure.

We do not intend to give the impression that the universe of a structure needs to be a set of numbers. Now we present an example of a structure which might look a little artificial at first, but this type of structure will play a crucial role in Section 2.4. Let F be a formula containing at least one constant (i.e. a function symbol with arity 0), and let $\mathcal{A} = (U_A, I_A)$ be a structure where U_A consists of all variable-free terms that can be built from the symbols occuring in F. For the example formula F above, we get

$$U_A = \{a, f(a), g(a, a), f(g(a, a)), g(f(a), a), \ldots\}.$$

The crucial point is the interpretation of function symbols. For the function symbol f in F and for any term $t \in U_A$, let $f^A(t)$ be the term $f(t) \in U_A$, and for the function symbol g in F and for any terms $t_1, t_2 \in U_A$ let $g^A(t_1, t_2)$ be the term $g(t_1, t_2) \in U_A$. Furthermore, let $a^A = a$. The reader should see what kind of interaction between syntax and semantics is going on here. The terms in U_A are interpreted by themselves. For a complete definition of a structure A, the interpretation of the predicate symbol P still has to be given (i.e. the definition of I_A has to be extended to P). We leave it to the reader to do this in such a way that F becomes true (resp. false) under A.

Definition (semantics of predicate logic – continued)

Let F be a formula and let $A = (U_A, I_A)$ be a suitable structure for F. For each term t occuring in F, we denote its *value* under the structure A as $A(t)$ and define it inductively as follows.

1. If t is a variable (i.e., $t = x$), then we let $A(t) = x^A$.

2. If t has the form $t = f(t_1, \ldots, t_k)$ where $t_1, \ldots t_k$ are terms and f is a function symbol of arity k, then we let $A(t) = f^A(A(t_1), \ldots, A(t_k))$.

The rule 2 also includes the possibility that f has arity 0, that is, t has the form $t = a$. In this case we get $A(t) = a^A$.

Similarly, we define the *(truth-)value* of the formula F, denoted $A(F)$, under the structure A by an inductive definition.

1. If F has the form $F = P(t_1, \ldots, t_k)$ where $t_1, \ldots, t_k$ are terms and P is a predicate symbol of arity k, then

$$A(F) = \begin{cases} 1, & \text{if } (A(t_1), \ldots, A(t_k)) \in P^A \\ 0, & \text{otherwise} \end{cases}$$

2. If F has the form $F = \neg G$, then

$$A(F) = \begin{cases} 1, & \text{if } A(G) = 0 \\ 0, & \text{otherwise} \end{cases}$$

3. If F has the form $F = (G \wedge H)$, then

$$A(F) = \begin{cases} 1, & \text{if } A(G) = 1 \text{ and } A(H) = 1 \\ 0, & \text{otherwise} \end{cases}$$

4. If F has the form $F = (G \vee H)$, then

$$A(F) = \begin{cases} 1, & \text{if } A(G) = 1 \text{ or } A(H) = 1 \\ 0, & \text{otherwise} \end{cases}$$

5. If F has the form $F = \forall x G$, then

$$A(F) = \begin{cases} 1, & \text{if for all } u \in U_A, A_{[x/u]}(G) = 1 \\ 0, & \text{otherwise} \end{cases}$$

Here, $A_{[x/u]}$ is the structure A', which is identical to A with the exception of the definition of $x^{A'}$: No matter whether I_A is defined on x or not, we let $x^{A'} = u$.

6. If F has the form $F = \exists x G$, then

$$A(F) = \begin{cases} 1, & \text{if there exists some } u \in U_A \text{ such that } A_{[x/u]}(G) = 1 \\ 0, & \text{otherwise} \end{cases}$$

If for a formula F and a suitable structure A we have $A(F) = 1$, then we denote this by $A \models F$ (we say, F is *true* in A, or A is a *model* for F). If every suitable structure for F is a model for F, then we denote this by $\models F$ (F is *valid*), otherwise $\not\models F$. If there is at least one model for the formula F then F is called *satisfiable*, and otherwise *unsatisfiable* (or *contradictory*).

Exercise 44: Consider the following formula

$$F = \forall x \exists y P(x, y, f(z)) \ .$$

Define a suitable structure $A = (U_A, I_A)$ for F which is a model for F, and another structure $B = (U_B, I_B)$ which is not a model for F.

Many notions from propositional logic, like "consequence" and "equivalence" can be translated directly into predicate logic. We will use these notions in the following without giving new definitions.

Remarks:

1. Analogously to propositional logic, it can be shown

F is valid if and only if $\neg F$ is unsatisfiable.

2. Predicate logic can be understood as an extension of propositional logic in the following sense. If all predicate symbols are required to have arity 0 (then there is no use for variables, quantifiers, and terms), essentially we get the formulas in propositional logic where the predicates P_i^0 play the role of the atomic formulas A_i in propositional logic.

   It even suffices not to use variables (and therefore also no quantifiers) such that predicate logic "degenerates" to propositional logic. Let

   $$F = (Q(a) \vee \neg R(f(b), c)) \wedge P(a, b)$$

   be a formula without variables (but with predicate symbols of arity greater than 0). By identifying different atomic formulas in F with different atomic formulas A_i of propositional logic, such as

   $$
   \begin{aligned}
   Q(a) &\longleftrightarrow A_1 \\
   R(f(b), c) &\longleftrightarrow A_2 \\
   P(a, b) &\longleftrightarrow A_3
   \end{aligned}
   $$

   we get

   $$F' = (A_1 \vee \neg A_2) \wedge A_3 \ .$$

   Obviously, a formula obtained like F' from F is satisfiable (or valid) if and only if F is satisfiable (or valid).

3. Observe that a formula without occurences of a quantifier (e.g. the matrix of a given formula) can be transformed into an equivalent formula in **CNF** or **DNF** where only the tools from propositional logic are needed.

4. Although predicate logic is expressionally more "powerful" than propositional logic (i.e. more statements in colloquial language can be expressed formally), it is not powerful enough to express every conceivable statement (e.g. in mathematics). We obtain an even stronger power if we allow also quantifications that range over predicate or function symbols, like

   $$F = \forall P \exists f \forall x P(f(x)) \ .$$

   This is a matter of the so-called *second order* predicate logic (that we will not study in this book). What we consider here is the *first order* predicate logic. The elements of the universe (symbolized by the variables in a formula) are understood as *first order* objects whereas predicates and functions defined on the universe are *second order* objects.

Exercise 45: Consider the following formulas F_1, F_2, F_3 which express that the predicate P is reflexive, symmetric and transitive.

$$
\begin{aligned}
F_1 &= \forall x P(x, x) \\
F_2 &= \forall x \forall y (P(x, y) \rightarrow P(y, x)) \\
F_3 &= \forall x \forall y \forall z ((P(x, y) \wedge P(y, z)) \rightarrow P(x, z))
\end{aligned}
$$

Show that none of these formulas is a consequence of the other two by presenting structures which are models for two of the formulas, but not for the respective third formula.

Exercise 46: In predicate logic with *identity* the symbol $=$ is also permitted in formulas (as a special binary predicate with a fixed interpretation) which is to be interpreted as identity (of values) between terms. How has the syntax (i.e. the definition of formulas) and the semantics (the definition of $\mathcal{A}(F)$) of predicate logic to be extended to obtain the predicate logic with identity?

Exercise 47: Which of the following structures are models for the formula

$$ F = \exists x \exists y \exists z (P(x, y) \wedge P(z, y) \wedge P(x, z) \wedge \neg P(z, x)) \ ? $$

(a) $U_\mathcal{A} = \mathbb{N}$, $P^\mathcal{A} = \{(m, n) \mid m, n \in \mathbb{N}, m < n\}$

(b) $U_\mathcal{A} = \mathbb{N}$, $P^\mathcal{A} = \{(m, m+1) \mid m \in \mathbb{N}\}$

(c) $U_\mathcal{A} = 2^\mathbb{N}$ (the power set of $\mathbb{N}$),
$P^\mathcal{A} = \{(A, B) \mid A, B \subseteq \mathbb{N}, A \subseteq B\}$

Exercise 48: Let F be a formula, and let $x_1, \ldots, x_n$ be the variables that occur free in F. Show:

(a) F is valid if and only if $\forall x_1 \forall x_2 \cdots \forall x_n F$ is valid,

(b) F is satisfiable if and only if $\exists x_1 \exists x_2 \cdots \exists x_n F$ is satisfiable.

Exercise 49: Find a closed satisfiable formula F, such that for every model $\mathcal{A} = (U_{\mathcal{A}}, I_{\mathcal{A}})$ of F, $|U_{\mathcal{A}}| \geq 3$.

Exercise 50: Let F be a satisfiable formula and let $\mathcal{A}$ be a model for F with $|U_{\mathcal{A}}| = n$. Show that for every $m \geq n$ there is a model $\mathcal{B}_m$ for F with $|U_{\mathcal{B}_m}| = m$. Furthermore, there is a model $\mathcal{B}_\infty$ for F with $|U_{\mathcal{B}_\infty}| = \infty$.

Hint: Pick some element u from $U_{\mathcal{A}}$, and add new elements to $U_{\mathcal{B}_m}$ having the same properties as u.

Exercise 51: Find a satisfiable formula F of predicate logic with identity such that for every model $\mathcal{A}$ of F, $|U_{\mathcal{A}}| \leq 2$.

This exercise seems to contradict the previous exercise. Convince yourself that there is no contradiction!

Exercise 52: Find formulas of predicate logic with identity (cf. Exercise 46) which contain a binary predicate symbol P (or a unary function symbol f) and which express:

(a) P is a anti-symmetric relation.

(b) f is a one-one function.

(c) f is a function which is onto.

Exercise 53: Formulate a satisfiable formula F in predicate logic with identity (cf. Exercise 46) in which a binary function symbol f occurs such that for every model $\mathcal{A}$ of F it holds:

$(U_\mathcal{A}, f^\mathcal{A})$ is a group.

Exercise 54: A *stack* is a well known abstract data structure in Computer Science. Certain predicates and functions (better: operations) are defined to test the status of the stack or to manipulate the stack. E.g., *IsEmpty* is a unary predicate expressing the fact that the stack is empty, and *nullstack* is a constant that stands for the empty stack. Further, *top* (giving the top element of the stack) and *pop* are unary functions, and *push* is a binary function (which gives the new stack after pushing a new element on top of the given stack).

"Axiomatize" these operations which are allowed on a stack by a formula in predicate logic in such a way that every model of this formula can be understood as an (abstract) stack.

Hint: A possible part of such a formula might be the formula

$$\forall x \forall y (top(push(x, y)) = x)$$

2.2 Normal Forms

The concept of (semantic) equivalence can be translated into predicate logic in the obvious way: two formulas F and G of predicate logic are equivalent (symbolically: $F \equiv G$) if for all structures $\mathcal{A}$ which are suitable for both F and G, $\mathcal{A}(F) = \mathcal{A}(G)$.

Also we observe that all equivalences which have been proved for formulas in propositional logic still hold in predicate logic, e.g. deMorgan's law:

$$\neg(F \wedge G) \equiv (\neg F \vee \neg G)$$

For the purpose of manipulating formulas of predicate logic, to convert them to certain normal forms etc., we need equivalences which also include quantifiers.

Theorem

Let F and G be arbitrary formulas.

1. $\neg\forall x F \equiv \exists x \neg F$
   $\neg\exists x F \equiv \forall x \neg F$

2. If x does not occur free in G, then
   $$\begin{aligned}
   (\forall x F \wedge G) &\equiv \forall x(F \wedge G) \\
   (\forall x F \vee G) &\equiv \forall x(F \vee G) \\
   (\exists x F \wedge G) &\equiv \exists x(F \wedge G) \\
   (\exists x F \vee G) &\equiv \exists x(F \vee G)
   \end{aligned}$$

3. $(\forall x F \wedge \forall x G) \equiv \forall x(F \wedge G)$
   $(\exists x F \vee \exists x G) \equiv \exists x(F \vee G)$

4. $\forall x \forall y F \equiv \forall y \forall x F$
   $\exists x \exists y F \equiv \exists y \exists x F$

Proof: As an example, we only present the proof for the first equivalence in 2. Let $\mathcal{A} = (U_\mathcal{A}, I_\mathcal{A})$ be a structure, suitable for both sides of the equivalence to be proved. Then we have:

$\mathcal{A}(\forall x F \wedge G) = 1$

iff $\mathcal{A}(\forall x F) = 1$ and $\mathcal{A}(G) = 1$

iff for all $u \in U_\mathcal{A}$, $\mathcal{A}_{[x/u]}(F) = 1$ and $\mathcal{A}(G) = 1$

iff for all $u \in U_\mathcal{A}$, $\mathcal{A}_{[x/u]}(F) = 1$ and $\mathcal{A}_{[x/u]}(G) = 1$ (because x is not free in G, we have $\mathcal{A}(G) = \mathcal{A}_{[x/u]}(G)$)

iff for all $u \in U_\mathcal{A}$, $\mathcal{A}_{[x/u]}((F \wedge G)) = 1$

iff $\mathcal{A}(\forall x(F \wedge G)) = 1$.

∎

It is even more interesting to observe which pairs of very similar looking formulas are *not* equivalent:

$$\begin{aligned}
(\forall x F \vee \forall x G) &\not\equiv \forall x(F \vee G) \\
(\exists x F \wedge \exists x G) &\not\equiv \exists x(F \wedge G)
\end{aligned}$$

Exercise 55: Confirm this by exhibiting counterexamples (i.e. structures which are models for one of the formulas, but not for the other).

Exercise 56: Show that $F = (\exists x P(x) \rightarrow P(y))$ is equivalent to $G = \forall x(P(x) \rightarrow P(y))$.

Exercise 57: Prove that $\forall x \exists y P(x, y)$ is a consequence of $\exists u \forall v P(v, u)$, but not vice versa.

We further observe that the substitution theorem from propositional logic analogously holds in predicate logic. The induction proof (on the formula structure) that was given in Section 1.2 can be extended to the cases that can occur for formulas of predicate logic (*Case 4: F* has the form $F = \exists x G$, *Case 5: F* has the form $F = \forall x G$).

This leads over to the next remark. Induction proofs on the formula structure can be done in predicate logic as well (with more cases). Since the (inductive) definition of terms precedes the definition of formulas, and terms are parts of formulas, it is sometimes necessary to prove the assertion (or an adaptation of the assertion) inductively for terms first, and then for formulas.

Observe that the equivalences 1–3 in the above theorem, applied from left to right, "drive the quantifiers in front of the formula".

Example:

$$(\neg(\exists x P(x, y) \vee \forall z Q(z)) \wedge \exists w P(f(a, w)))$$
$$\equiv \quad ((\neg\exists x P(x, y) \wedge \neg\forall z Q(z)) \wedge \exists w P(f(a, w))) \text{ (de Morgan)}$$
$$\equiv \quad ((\forall x \neg P(x, y) \wedge \exists z \neg Q(z)) \wedge \exists w P(f(a, w)) \text{ (by 1.)}$$
$$\equiv \quad (\exists w P(f(a, w)) \wedge (\forall x \neg P(x, y) \wedge \exists z \neg Q(z))) \text{ (commutativity)}$$
$$\equiv \quad \exists w(P(f(a, w)) \wedge \forall x(\neg P(x, y) \wedge \exists z \neg Q(z))) \text{ (by 2.)}$$
$$\equiv \quad \exists w(\forall x(\exists z \neg Q(z) \wedge \neg P(x, y)) \wedge P(f(a, w))) \text{ (commutativity)}$$
$$\equiv \quad \exists w(\forall x \exists z(\neg Q(z) \wedge \neg P(x, y)) \wedge P(f(a, w))) \text{ (by 2.)}$$
$$\equiv \quad \exists w \forall x \exists z(\neg Q(z) \wedge \neg P(x, y) \wedge P(f(a, w))) \text{ (by 2.)}$$

Several points need to be observed. The order of the quantifiers which results at the end, is not necessarily uniquely determined from the beginning. Actually, it depends on the type and the order of the applied equivalences. In the above example, every permutation of "$\exists w$", "$\forall x$" and "$\exists z$" would have been achievable. (It is not always like this). But adjacent quantifiers of the same type can always be swapped (see 4.).

To make it possible that the equivalences under 2. can always be applied, we need to rename variables (in such that way that we get an equivalent formula).

Definition (substitution)

Let F be a formula, x a variable, and t a term. Then, $F[x/t]$ denotes the formula, obtained from F by substituting t for every free occurence of x in F.

By $[x/t]$, a *substitution* is described. In the following, we treat substitutions as independent objects, describing a mapping from the set of formulas to the set of formulas. Such substitutions can be concatenated, e.g.

$$sub = [x/t_1][y/t_2]$$

describes the effect of first substituting in a formula all free occurences of x by t_1, and then, all free occurences of y by t_2. (Note that t_1 can contain occurences of y).

Exercise 58: Prove by induction on the formula structure the following *translation lemma*. Here, t is a variable-free term.

$$\mathcal{A}(F[x/t]) = \mathcal{A}_{[x/\mathcal{A}(t)]}(F)$$

The proof of the following lemma is just as easy.

Lemma (renaming of bound variables)

Let $F = QxG$ be a formula where $Q \in \{\exists, \forall\}$. Let y be a variable that does not occur free in G. Then, $F \equiv QyG[x/y]$.

By systematic applications of the previous lemma where always new variables have to be taken for y, the following lemma can be proved. Call a formula *rectified* if no variable occurs both bound and free and if all quantifiers in the formula refer to different variables.

Lemma

For every formula F there is an equivalent formula G in rectified form.

Exercise 59: Find an equivalent and rectified formula for

$$F = \forall x \exists y P(x, f(y)) \land \forall y (Q(x, y) \lor R(x)).$$

The above example already shows that every formula can be transformed into an equivalent and rectified formula where all quantifiers stand "in front". We summarize this situation more formally in the following definition and theorem.

Definition (prenex form)

A formula is in *prenex form* if it has the form

$$Q_1 y_1 Q_2 y_2 \ldots Q_n y_n F,$$

where $Q_i \in \{\exists, \forall\}, n \geq 0$, and the y_i are variables. Further, F does not contain a quantifier.

Theorem

For every formula F there exists an equivalent (and rectified) formula G in prenex form.

Proof (by induction on the formula structure of F):
If F is an atomic formula, then F already has the desired form. Thus we choose $G = F$.

For the induction step we consider the different cases.

1. Let F have the form $\neg F_1$ and $G_1 = Q_1 y_1 Q_2 y_2 \cdots Q_n y_n G'$ is the formula, equivalent to F_1, which exists by induction hypothesis. Then we have

$$F \equiv \overline{Q_1} y_1 \overline{Q_2} y_2 \cdots \overline{Q_k} y_k \neg G'$$

where $\overline{Q_i} = \exists$ if $Q_i = \forall$, and $\overline{Q_i} = \forall$ if $Q_i = \exists$. This formula has the desired form.

2. Let F have the form $(F_1 \circ F_2)$ where $\circ \in \{\wedge, \vee\}$, then there are, by induction hypothesis, equivalent formulas G_1, G_2 in prenex form for F_1 and F_2, resp. By renaming the bound variables, say in G_1, we can make the bound variables of G_1 and G_2 disjoint. Let then G_1 have the form $Q_1 y_1 Q_2 y_2 \cdots Q_k y_k G_1'$ and G_2 have the form $Q_1' z_1 Q_2' z_2 \cdots Q_l' z_l G_2'$ where $Q_i, Q_j' \in \{\exists, \forall\}$. It follows that F is equivalent to

$$Q_1 y_1 Q_2 y_2 \cdots Q_k y_k Q_1' z_1 Q_2' z_2 \cdots Q_l' z_l (G_1' \circ G_2')$$

This formula has the desired rectified prenex form.

3. If F has the form $Q x F_1$ where $Q \in \{\exists, \forall\}$, then the formula F_1 is equivalent, by induction hypothesis, to a formula of the form

$$Q_1 y_1 Q_2 y_2 \cdots Q_k y_k F_1'.$$

By renaming bound variables, we can assume that the variable x is different from all the variables y_i. Then, F is equivalent to

$$Q x Q_1 y_1 Q_2 y_2 \cdots Q_k y_k F_1'.$$

∎

Exercise 60: Implicit in the above proof, there is an algorithm hidden to convert formulas into rectified prenex form. Formulate such an algorithm in a more direct way, using a PASCAL-like notation.

Exercise 61: Convert the formula

$$F = (\forall x \exists y P(x, g(y, f(x)))) \vee \neg Q(z)) \vee \neg \forall x R(x, y)$$

into rectified prenex form.

From now on, we use the abbreviation **RPF** for "rectified and in prenex form".

Definition (Skolem form)

For each formula F in **RPF** we define its *Skolem formula* as the result of applying the following algorithm to F.

> **while** F contains an existential quantifier **do**
> > **begin**
> > > Let F have the form $F = \forall y_1 \forall y_2 \cdots \forall y_n \exists z G$ for some formula G in **RPF** and $n \geq 0$ (the block of universal quantifiers could also be empty);
> > >
> > > Let f be a new function symbol of arity n that does not yet occur in F ;
> > >
> > > $F := \forall y_1 \forall y_2 \cdots \forall y_n G[z/f(y_1, y_2, \ldots, y_n)]$;
> > >
> > > (i.e. the existential quantifier in F is canceled and each occurence of the variable z in G is substituted by $f(y_1, y_2, \ldots, y_n)$)
> >
> > **end.**

Exercise 62: Find the Skolem form of the formula

$$\forall x \exists y \forall z \exists w (\neg P(a, w) \vee Q(f(x), y)).$$

Theorem

For each formula F in **RPF**, F is satisfiable if and only if the Skolem form of F is satisfiable.

Proof: We show that after each application of the **while**-loop a formula results which is satisfiable if and only if the original formula is satisfiable. Hence, let

$$F = \forall y_1 \forall y_2 \cdots \forall y_n \exists z G .$$

After one application of the **while**-loop we obtain the formula

$$F' = \forall y_1 \forall y_2 \cdots \forall y_n G[z/f(y_1, y_2, \ldots, y_n)] .$$

Let us suppose first that F' is satisfiable. That is, there is a structure $\mathcal{A}$, suitable for F', with $\mathcal{A}(F') = 1$. Then $\mathcal{A}$ is also suitable for F, and we get

$$\text{for all } u_1, u_2, \ldots, u_n \in U_{\mathcal{A}} \text{ ,}$$
$$\mathcal{A}_{[y_1/u_1][y_2/u_2]\cdots[y_n/u_n]}(G[z/f(y_1, y_2, \cdots y_n)]) = 1 \text{ .}$$

By the translation lemma,

$$\text{for all } u_1, u_2, \ldots, u_n \in U_{\mathcal{A}} \text{ ,}$$
$$\mathcal{A}_{[y_1/u_1][y_2/u_2]\cdots[y_n/u_n][z/v]}(G) = 1 \text{ ,}$$

where $v = f^{\mathcal{A}}(u_1, u_2, \cdots, u_n)$. Hence we get

$$\text{for all } u_1, u_2, \ldots, u_n \in U_{\mathcal{A}} \text{ there exists a } v \in U_{\mathcal{A}} \text{ such that}$$
$$\mathcal{A}_{[y_1/u_1][y_2/u_2]\cdots[y_n/u_n][z/v]}(G) = 1 \text{ .}$$

Therefore,

$$\mathcal{A}(\forall y_1 \forall y_2 \cdots \forall y_n \exists z G) = 1 \text{ .}$$

In other words, $\mathcal{A}$ is also a model for F.

Conversely, suppose F has the model $\mathcal{A} = (U_{\mathcal{A}}, I_{\mathcal{A}})$. We can assume that $I_{\mathcal{A}}$ is undefined on function symbols that do not occur in F. Hence, $I_{\mathcal{A}}$ is not defined on f and not (yet) suitable for F'. Since $\mathcal{A}(F) = 1$, we have

$$\text{for all } u_1, u_2, \ldots, u_n \in U_{\mathcal{A}} \text{ there exists a } v \in U_{\mathcal{A}} \qquad (*)$$
$$\text{such that } \mathcal{A}_{[y_1/u_1]\cdots[y_n/u_n][z/v]}(G) = 1 \text{ .}$$

Now we define a new structure $\mathcal{A}'$ which is an extension of $\mathcal{A}$ such that $I_{\mathcal{A}'}$ is additionally defined on f. We let $f^{\mathcal{A}'}$ be defined as

$$f^{\mathcal{A}'}(u_1, \ldots, u_n) = v,$$

where $v \in U_{\mathcal{A}} = U_{\mathcal{A}'}$ is chosen according to $(*)$. (At this point of the proof, the *axiom of choice* is used which guarantees the existence of such a "non-constructively" defined function). Using this definition of $f^{\mathcal{A}'}$, we obtain

$$\text{for all } u_1, \ldots, u_n \in U_{\mathcal{A}} \text{ ,}$$
$$\mathcal{A}'_{[y_1/u_1]\cdots[y_n/u_n][z/f^{\mathcal{A}'}(u_1,\ldots,u_n)]}(G) = 1.$$

Using the translation lemma,

$$\text{for all } u_1, \ldots, u_n \in U_{\mathcal{A}} \ ,$$
$$\mathcal{A}'_{[y_1/u_1]\cdots[y_n/u_n]}(G[z/f(y_1, \ldots, y_n)]) = 1,$$

and therefore,

$$\mathcal{A}'(\forall y_1 \cdots \forall y_n G[z/f(y_1, \ldots, y_n)]) = 1.$$

Hence, $\mathcal{A}'$ is a model for F'. ∎

Notice that the transformation of a formula to Skolem form does not preserve equivalence (because of the new function symbol(s) occuring in the Skolem formula). What we have shown is a weaker type of equivalence with respect to satisfiability: F is satisfiable if and only if F' is satisfiable. In the following, we call this situation *s-equivalence*.

Exercise 63: Apply all transformational steps introduced in this chapter (rectification, prenex form, Skolem form) to the formula

$$\forall z \exists y (P(x, g(y), z) \lor \neg \forall x Q(x)) \land \neg \forall z \exists x \neg R(f(x, z), z).$$

Exercise 64: If we modify the algorithm to produce the Skolem form such that the roles of $\forall$ and $\exists$ are swapped, then we obtain an algorithm which transforms a formula F in **RPF** into a formula F' with no occurences of universal quantifiers. Prove that F is valid if and only if F' is valid.

Exercise 65: Construct an algorithm that produces a Skolem form of a rectified formula *directly*, i.e. without the intermediate step of producing a prenex form.

Hint: It is important to distinguish between existential (universal) quantifiers in the original formula that lie within the "scope" of an even (odd, resp.) number of negation signs.

Finally, we want to summarize all the transformations which should be applied to a general formula to obtain an s-equivalent formula which is in appropriate form for the various algorithms considered in the next sections.

Given : A formula F in predicate logic (with possible occurences of free variables).

1. Rectify F by systematic renaming of bound veriables. The result is a formula F_1 equivalent to F.

2. Let $y_1, \ldots, y_n$ be the variables that occur free in F_1. Substitute F_1 by $F_2 = \exists y_1 \exists y_2 \ldots \exists y_n F_1$. Then, F_2 is s-equivalent to F_1 (cf. Exercise 48) and also to F. Further, F_2 is closed.

3. Produce from F_2 a formula F_3 in prenex form. F_3 is equivalent to F_2, hence s-equivalent to F.

4. Eliminate the existential quantifiers in F_3 by transforming F_3 into a Skolem formula F_4. The formula F_4 is s-equivalent to F_3, hence s-equivalent to F.

5. Convert the matrix of F_4 into **CNF** (and write the resulting formula F_5 down as a set of clauses).

We demonstrate the above procedure with an example. Let

$$F = (\neg \exists x (P(x,z) \vee \forall y Q(x, f(y)))) \vee \forall y P(g(x,y), z))$$

be given. Renaming y to w in the second disjunct gives a rectified form

$$F_1 = (\neg \exists x (P(x,z) \vee \forall y Q(x, f(y)))) \vee \forall w P(g(x,w), z))$$

The variable z occurs free in F_1. Hence we let

$$F_2 = \exists z ((\neg \exists x (P(x,z) \vee \forall y Q(x, f(y)))) \vee \forall w P(g(x,w), z))).$$

Converting to prenex form gives (for example)

$$F_3 = \exists z \forall x \exists y \forall w ((\neg (P(x,z) \wedge \neg Q(x, f(y)))) \vee P(g(x,w), z)).$$

Now we produce the Skolem form. A new function symbol a of arity 0 (i.e. a constant) is substituted for z and $h(x)$ is substituted for y.

$$F_4 = \forall x \forall w ((\neg (P(x,a) \wedge \neg Q(x, f(h(x))))) \vee P(g(x,w), a)).$$

Transforming the matrix of F_4 into **CNF** yields

$$F_5 = \forall x \forall w ((\neg (P(x,a) \vee P(g(x,w), a)) \wedge (\neg Q(x, f(h(x)))) \vee P(g(x,w), a)).$$

Now, F_5 can be written as a clause set:

$$\{\{\neg(P(x, a), P(g(x, w), a)\}, \{\neg Q(x, f(h(x)))), P(g(x, w), a)\}\}.$$

Every variable is understood as universally bounded. Hence we do not need to write down the universal quantifiers explicitly.

This clause presentation of formulas in predicate logic is the starting point for several algorithms, based on resolution, to be presented in Sections 2.5 and 2.6.

Finally we remark that all the transformational steps can be done algorithmically.

2.3 Undecidability

A general theme of this book is the search for an algorithmic test for satisfiability or validity of formulas. We will see in this section that general algorithms of this type cannot exist for formulas in predicate logic. Briefly, predicate logic is *undecidable*. (More precisely, the satisfiability problem and the validity problem for formulas in predicate logic are undecidable). We must be content with so-called *semi-decision algorithms* which will be presented in the next section.

The truth table method for testing satisfiability or validity of formulas discussed in the chapter on propositional logic could be derived from the insight that it is enough to test a finite (although exponential) number of truth assignments. In predicate logic we have to deal with structures instead of truth assignments. The question is whether we can restrict our attention to a selection of finitely many structures, and also, to structures of finite size. As already suggested, this kind of direct adaption of the truth table method does not work.

Observation: There exist formulas in predicate logic which are satisfiable, but have no models of finite size (i.e. with a finite universe).

Consider the formula

$$\begin{aligned} F \quad &= \quad \forall x P(x, f(x)) \\ &\wedge \forall y \neg P(y, y) \\ &\wedge \forall u \forall v \forall w ((P(u, v) \wedge P(v, w)) \rightarrow P(u, w)). \end{aligned}$$

This formula F is satisfiable, because it has for example the following model $\mathcal{A} = (U_{\mathcal{A}}, I_{\mathcal{A}})$ where

$$
\begin{aligned}
U_{\mathcal{A}} &= \{0, 1, 2, 3, \ldots\} = \mathbb{N} \\
P^{\mathcal{A}} &= \{(m, n) \mid m < n\}, \\
f^{\mathcal{A}}(n) &= n + 1.
\end{aligned}
$$

But this formula does not possess a finite model. Suppose, $\mathcal{B} = (U_{\mathcal{B}}, I_{\mathcal{B}})$ is such a model for F. Then let u be an arbitrary element of $U_{\mathcal{B}}$. Consider the sequence

$$ u_0, u_1, u_2, \ldots \in U_{\mathcal{B}} \text{ where } u_0 = u \text{ and } u_{i+1} = f^{\mathcal{B}}(u_i). $$

Since $U_{\mathcal{B}}$ is finite, there exist natural numbers i and j, $i < j$, such that $u_i = u_j$. By the first subformula of F we have:

$$ (u_0, u_1) \in P^{\mathcal{B}}, (u_1, u_2) \in P^{\mathcal{B}}, (u_2, u_3) \in P^{\mathcal{B}}, \ldots $$

Further, the third subformula of F says that $P^{\mathcal{B}}$ must be a transitive relation. This implies that $(u_i, u_j) \in P^{\mathcal{B}}$. Since $u_i = u_j$, we have found an element v of the universe $U_{\mathcal{B}}$ with $(v, v) \in P^{\mathcal{B}}$. But this contradicts the second subformula of F which says that $P^{\mathcal{B}}$ must be non-reflexive. This shows that F has only infinite models.

It should be said that the above argument is not yet a formal proof of undecidability of predicate logic. The existence of satisfiable formulas which have only infinite models just shows that there is no direct translation of the truth table method into predicate logic to yield a decision procedure. The possible existence of totally different algorithms is not touched by the above argument.

For a formal presentation of an undecidability proof, it is necessary to clarify and formally define the notions "computation" and "algorithm" first. After all, we need to show that there is no algorithm that is able to compute (in a finite amount of time) whether a given formula is, say, satisfiable. These issues are subject of a different field, *computability theory*, which is not the subject of this book (see, for example, the books by Manna or Hopcroft and Ullman). Therefore we proceed with some informal explanations, and then use a result from computability theory, namely that a specific well known problem is undecidable. Relying on this fact, we can proceed formally.

In computability theory, a function is called *computable* (or a problem is called *decidable*) if there is an abstract mathematical machine (Turing-machine) which, started with an input which is in the function domain (which is a syntactically correct instance for the problem, resp.) halts after a finite number of steps and outputs the correct function value (answers correctly "yes" or "no", according to the problem definition). If no such machine exists, then the function (problem) is called non-computable (undecidable).

We have to deal with *problems* in the following. Such a problem is given by specifying the form of a syntactically correct instance for the problem, and what the question to be solved is.

In particular, we will show that the following problem is undecidable.

Instance: A formula F in predicate logic.

Question: Is F valid?

In what follows, we use a result from computability theory: the following problem, called *Post's Correspondence Problem* (PCP for short), is undecidable (see Hopcroft and Ullman).

Instance: A finite sequence $(x_1, y_1), \ldots, (x_k, y_k)$ of pairs of non-empty strings over the alphabet $\{0, 1\}$.

Question: Does there exist a finite sequence of indices $i_1, i_2, \ldots, i_n \in \{1, \ldots, k\}$, $n \geq 1$, such that $x_{i_1} x_{i_2} \ldots x_{i_n} = y_{i_1} y_{i_2} \ldots y_{i_n}$?

In the case that $i_1, \ldots, i_n$ exists, we call it a *solution* of the PCP.

Example: The correspondence problem for

$$\mathbf{K} = ((1, 101), (10, 00), (011, 11)),$$

that is

$$x_1 = 1 \quad x_2 = 10 \quad x_3 = 011$$
$$y_1 = 101 \quad y_2 = 00 \quad y_3 = 11$$

has the solution (1,3,2,3) because:

$$x_1 x_3 x_2 x_3 = 101110011 = y_1 y_3 y_2 y_3$$

Exercise 66: Show that the following instance of PCP has a solution:

$$x_1 = 001 \quad x_2 = 01 \quad x_3 = 01 \quad x_4 = 10$$
$$y_1 = 0 \quad y_2 = 011 \quad y_3 = 101 \quad y_4 = 001.$$

(Warning: the shortest solution consists of 66 indices. Without using a computer, the solution can be found if constructed "from behind").

We use the proof method of *reduction* to show that the validity problem is undecidable. That is, from a hypothetical decision algorithm for the validity problem we derive the existence of a decision algorithm for the PCP – which is in contradiction to the result stated above. Hence, a decision algorithm for the validity problem does not exist, this means that the problem is undecidable.

Many known undecidabilty results have been shown by reduction. Also, it is very common to use the undecidability of the PCP – in particular, for undecidability proofs in Formal Language Theory.

Theorem (Church)

The validity problem for formulas of predicate logic is undecidable.

Proof: As discussed above, the task is to define an algorithmic method that transforms arbitrary instances K for the PCP into certain instances, i.e. formulas, $F = F_K$ for the validity problem, such that K has a solution if and only if the formula F_K is valid. If this can be shown then the hypothetical existence of a decision algorithm for the validity problem implies the existence of a decision algorithm for the PCP. Hence, let

$$K = ((x_1, y_1), (x_2, y_2), \ldots, (x_k, y_k))$$

be an arbitrary correspondence problem. The desired formula $F = F_K$ contains a constant a and two unary function symbols f_0, f_1. Furthermore, a binary predicate symbol P occurs in F. For a more succinct representation of the formula, we use the following abbreviation. Instead of

$$f_{j_1}(f_{j_2}(\ldots f_{j_l}(x)\ldots)) \text{ with } j_i \in \{0, 1\}$$

we write

$$f_{j_l \ldots j_2 j_1}(x).$$

(The indices now stand in reverse order).

Our formula $F = F_K$ has the form

$$F = ((F_1 \wedge F_2) \rightarrow F_3).$$

The subformulas are

$$F_1 = \bigwedge_{i=1}^{k} P(f_{x_i}(a), f_{y_i}(a))$$

$$F_2 = \forall u \forall v (P(u,v) \rightarrow \bigwedge_{i=1}^{k} P(f_{x_i}(u), f_{y_i}(v)))$$

$$F_3 = \exists z P(z,z).$$

Obviously, for given **K**, *F* can be computed from **K** in a finite amount of time. We have to show that the formula *F* is valid if and only if the correspondence problem **K** has a solution.

Let us assume first that *F* is valid. Then every suitable structure for *F* is a model. In particular, the following structure $\mathcal{A} = (U_{\mathcal{A}}, I_{\mathcal{A}})$ must be a model for *F*.

$$
\begin{aligned}
U_{\mathcal{A}} &= \{0,1\}^*, \\
a^{\mathcal{A}} &= \varepsilon \text{ (the empty string)}, \\
f_0^{\mathcal{A}}(\alpha) &= \alpha 0 \text{ (the concatenation of } \alpha \text{ and } 0), \\
f_1^{\mathcal{A}}(\alpha) &= \alpha 1 \text{ (the concatenation of } \alpha \text{ and } 1), \\
P^{\mathcal{A}} &= \{(\alpha,\beta) \mid \alpha, \beta \in \{0,1\}^+ \text{ and there are indices} \\
&\quad i_1, i_2, \ldots, i_t \text{ such that } \alpha = x_{i_1} x_{i_2} \ldots x_{i_t} \text{ and} \\
&\quad \beta = y_{i_1} y_{i_2} \ldots y_{i_t} \}.
\end{aligned}
$$

That is, a pair of strings (α, β) is in $P^{\mathcal{A}}$ if α can be built up from the x_i by the same sequence of indices as β from the y_i. It is easily seen that $\mathcal{A}$ is suitable for *F*. Hence $\mathcal{A} \models F$. Further, it can be checked that $\mathcal{A} \models F_1$ and $\mathcal{A} \models F_2$. Since *F* has the form of an implication $((F_1 \wedge F_2) \rightarrow F_3)$, it follows that $\mathcal{A} \models F_3$. This means that there exists some α such that $(\alpha, \alpha) \in P^{\mathcal{A}}$. Hence **K** has a solution.

Conversely, suppose that **K** has the solution $i_1, i_2, \ldots, i_n$. Let $\mathcal{A}$ be an arbitrary structure suitable for *F*. We have to show that $\mathcal{A} \models F$. If $\mathcal{A} \not\models F_1$ or $\mathcal{A} \not\models F_2$, then, by the form of *F*, $\mathcal{A} \models F$ follows immediately. Hence let us assume that $\mathcal{A} \models F_1$ and $\mathcal{A} \models F_2$, thus $\mathcal{A} \models (F_1 \wedge F_2)$. We now define a mapping (an *embedding*) $\mu : \{0,1\}^* \rightarrow U_{\mathcal{A}}$ by $\mu(\varepsilon) = a^{\mathcal{A}}$ and $\mu(x) = \mathcal{A}(f_x(a))$ for $x \neq \varepsilon$.

Because $\mathcal{A} \models F_1$, we have for $i = 1, 2, \ldots, k$: $(\mu(x_i), \mu(y_i)) \in P^{\mathcal{A}}$. Because of $\mathcal{A} \models F_2$, we have for $i = 1, 2, \ldots, k$, that $(\mu(u), \mu(v)) \in P^{\mathcal{A}}$ implies $(\mu(ux_i), \mu(vy_i)) \in P^{\mathcal{A}}$. By induction, it follows that

$$(\mu(x_{i_1} x_{i_2} \ldots x_{i_n}), \mu(y_{i_1} y_{i_2} \ldots y_{i_n})) \in P^{\mathcal{A}}.$$

In other words, for $u = \mu(x_{i_1} x_{i_2} \ldots x_{i_n}) = \mu(y_{i_1} y_{i_2} \ldots y_{i_n})$ it is true that $(u, u) \in P^{\mathcal{A}}$. From this, we get $\mathcal{A} \models \exists z P(z, z)$, that is, $\mathcal{A} \models F_3$, and therefore, $\mathcal{A} \models F$.   ∎

Corollary

The satisfiability problem of predicate logic

> *Instance:* A formula F of predicate logic.
>
> *Question:* Is F satisfiable?

is undecidable.

Proof: A formula F is valid if and only if $\neg F$ is unsatisfiable. Therefore, the hypothetical existence of a decision algorithm for the satisfiability problem leads to a decision algorithm for the validity problem, and we have shown above that such an algorithm does not exist.   ∎

The reader will have noticed that the proof of this corollary is another example of the reduction method.

Exercise 67: Prove that the validity problem (and therefore also the satisfiability problem) is undecidable even for formulas without occurences of function symbols.

Exercise 68: Prove that the following variation of the PCP is decidable:

> *Instance:* A finite sequence of pairs $(x_1, y_1), \ldots, (x_k, y_k)$ where $x_i, y_i \in \{0, 1\}^+$.
>
> *Question:* Do there exist finite sequences of indices $i_1, i_2, \ldots, i_n$, $n \geq 1$, and $j_1, j_2, \ldots, j_m$, $m \geq 1$, such that $x_{i_1} x_{i_2} \ldots x_{i_n} = y_{j_1} y_{j_2} \ldots y_{j_m}$?

Exercise 69: In *monadic* predicate logic all the predicate symbols are unary (i.e. monadic) and no occurences of function symbols are allowed.

Prove: If some closed formula F of monadic predicate logic with the unary predicate symbols $P_1, \ldots, P_n$ is satisfiable, then there is already a model of cardinality 2^n. From this, conclude that satisfiability (and also validity) for formulas in monadic predicate logic is decidable.

Hint: Show that the universe of every model $\mathcal{A} = (U_{\mathcal{A}}, I_{\mathcal{A}})$ for F can be partitioned into at most 2^n equivalence classes where two elements $u, v \in U_{\mathcal{A}}$ are equivalent if they have the same truth value under each of $P_1^{\mathcal{A}}, \ldots, P_n^{\mathcal{A}}$. Then, a new model $\mathcal{B}$ can be defined for F whose universe consists of these equivalence classes.

Exercise 70: Show that the following problem is undecidable:

Instance: The description of an algorithm A.

Question: If A is started with its own description as input, does A stop?

Excursion (mathematical theories)

At this point, some important notions in Formal Logic shall be discussed. What is a formal mathematical theory? These issues play an important role in standard presentations of logic, but in this book with its emphasis on Computer Science and algorithmic aspects of logic, it is more a fringe area.

A *theory* is a non-empty set **T** of formulas – very often restricted to formulas obeying certain syntactical restrictions (e.g. only a given finite set of function symbols or predicate symbols may be allowed) – which is closed under consequence. More precisely, **T** is a theory, if for all $F_1, F_2, \ldots, F_n \in$ **T** and formulas G, if G is a consequence of $F_1, F_2, \ldots, F_n$ then $G \in$ **T**. The formulas which are elements of a theory **T** are called *theorems* of **T**.

Every theory **T** necessarily has to include all valid formulas (possibly only those obeying the syntactical restriction as above). Furthermore, a theory either contains *all* formulas, or it is disjoint from the set of unsatisfiable formulas. The former situation is the denerate case of an inconsistent theory. A theory is called *inconsistent* if it contains some closed formula F together with its negation $\neg F$. The following diagram indicates the situation of a non-degenerate theory **T**.

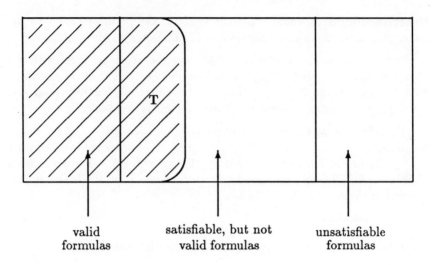

valid   satisfiable, but not   unsatisfiable
formulas   valid formulas   formulas

There are two different methods to define a particular theory.

The *model theoretic* method is to define a structure $\mathcal{A}$ first, and then take the theory of $\mathcal{A}$ (in symbols: $Th(\mathcal{A})$) as the set of all formulas for which $\mathcal{A}$ is a model. That is,

$$Th(\mathcal{A}) = \{F \mid \mathcal{A} \models F\}.$$

It is clear that a set of formulas of the form $Th(\mathcal{A})$ is really a theory, i.e. it is necessarily closed under consequence. Such a theory is automatically consistent. Further, such a theory is always *complete*, which means that for every closed formula F, either $F \in \mathbf{T}$ or $\neg F \in \mathbf{T}$ holds (but not both).

Examples for such model theoretically defined theories are $Th(\mathbb{N}, +)$ and $Th(\mathbb{N}, +, *)$. Here, $(\mathbb{N}, +)$ and $(\mathbb{N}, +, *)$ are the structures obtained by taking as universe $\mathbb{N}$ and interpretation of $+$ as usual addition and $*$ as usual multiplication. These theories are called *Presburger arithmethic* and (full) *arithmetic*, respectively. The formulas of the theories are restricted to consist of the function symbols $+$ and $*$ (and possibly further constant symbols and identity) only. For example,

$$\forall x \forall y \left((x + y) * (x + y) = (x * x) + (2 * x * y) + (y * y) \right)$$

is an element of $Th(\mathbb{N}, +, *)$.

The *axiomatic* method is to define a set of formulas $\mathbf{M}$, the *axioms*, and then take as the theory associated with $\mathbf{M}$ the set of formulas which are consequences of M. Formally,

$$Cons(\mathbf{M}) \quad = \quad \{G \mid \text{there are formulas } F_1, \ldots, F_n \in \mathbf{M},$$
$$\text{such that } G \text{ is a consequence of } \{F_1, \ldots, F_n\} \,\}.$$

Again, the formulas in $Cons(\mathbf{M})$ can be restricted to consist only of symbols which occur in M. It is required that such an axiom set M is decidable, i.e. for every formula F it should be possible to decide whether $F \in \mathbf{M}$ or not. In particular this is the case if M is finite.

A theory T is called *(finitely) axiomatizable* if there exists a (finite) axiom set M such that $\mathbf{T} = Cons(\mathbf{M})$. For example, the set of valid formulas of predicate logic is finitely axiomatizable, because

$$Cons(\emptyset) = \{F \mid F \text{ is valid }\}.$$

Another example is the *theory of groups*. This is $Cons(\mathbf{M})$ where

$$\mathbf{M} \quad = \quad \{\forall x \forall y \forall z(f(f(x,y),z) = f(x,f(y,z))),$$
$$\forall x(f(x,e) = x),$$
$$\forall x \exists y(f(x,y) = e)\}.$$

It can be shown that any axiomatizable theory is *semi-decidable* (which is the same as *recursively enumerable*; for an explanation of these notions see Section 2.4). Furthermore, every complete and axiomatizable theory is decidable. Now there are two main questions that can be investigated.

1. Are certain (axiomatizable) theories decidable? For example, we have seen in Section 2.3 that the finitely axiomatizable theory $Cons(\emptyset)$ is undecidable (and hence, it cannot be complete).

2. Are certain model theoretic theories axiomatizable – or even decidable? It can be shown that $Th(\mathbb{N}, +, *)$ is not axiomatizable (hence not decidable). In other words, every arithmetically correct axiom system M (for example: Peano arithmetic) necessarily is incomplete:

$$Cons(\mathbf{M}) \neq Th(\mathbb{N}, +, *).$$

Arithmetical correctness means that M (and therefore also $Cons(\mathbf{M})$) is included in $Th(\mathbb{N}, *, +)$. (This is Gödel's famous *incompleteness theorem*). This is in contrast to the fact that $Th(\mathbb{N}, +)$ is decidable (and therefore axiomatizable).

Exercise 71: Why is every complete and axiomatizable theory decidable?

2.4   Herbrand's Theory

One problem with dealing with formulas in predicate logic is that the definition of structures allows arbitrary sets as possible universes. It seems that there is no systematic way to find out the "inner structure" and cardinality of a potential model of a given formula. Can one enumerate all potential structures to test them for being a model? If so, how?

Indeed, in the last section it was shown that the problem of determining whether a given formula has a model or not is undecidable. This indicates a borderline which we will not be able to pass: We cannot expect to devise a decision algorithm. Nevertheless, in this section we will investigate the remaining positive aspects, insofar as they are not in contradiction to the undecidability result of the last section.

The (algorithmic) search for potential models of a formula can be restricted to certain *canonical* structures. This theory which we will develop in the following goes back to the work of Jacques Herbrand, Kurt Gödel and Thoralf Skolem. In particular, Herbrand's work is important for the approach taken here.

The starting point of our investigations are *closed* formulas, i.e. formulas without occurences of free variables, which are in Skolem form (hence also in **RPF**). In section 2.3 it was shown how every formula of predicate logic can be transformed into a s-equivalent formula of this kind.

Definition (Herbrand universe)

The *Herbrand universe* $D(F)$ of a closed formula F in Skolem form is the set of all variable-free terms that can be built from the components of F. In the special case that F does not contain a constant, we first choose an arbitrary constant, say a, and then build up the variable-free terms. More precisely, $D(F)$ is defined inductively as follows.

1. Every constant occuring in F is in $D(F)$. If F does not contain a constant, then a is in $D(F)$.

2. For every k-ary function symbol f that occurs in F, and for all terms $t_1, t_2, \ldots, t_k$ already in $D(F)$, the term $f(t_2, t_2, \ldots, t_k)$ is in $D(F)$.

Example: Consider the following formulas F and G.

$$F \;=\; \forall x \forall y \forall z P(x, f(y), g(z, x))$$

$$G \;=\; \forall x \forall y Q(c, f(x), h(y, b))$$

The formula F does not contain a constant. Therefore we get

$$D(F) \;=\; \{a, f(a), g(a, a), f(g(a, a)), f(f(a)), g(a, f(a)), g(f(a), a),$$
$$g(f(a), f(a)), \ldots\}$$

and

$$D(G) \;=\; \{c, b, f(c), f(b), h(c, c), h(c, b), h(b, c), h(b, b),$$
$$f(f(c)), f(f(b)), f(h(c, c)), f(h(c, b)), f(h(b, c)), \ldots\}$$

In the following, for a given formula F, $D(F)$ will be used as the "standard" universe to search for potential models for F – and we will show that this results in no loss in generality.

Definition (Herbrand structures)

Let F be a closed formula in Skolem form. Then every structure $\mathcal{A} = (U_{\mathcal{A}}, I_{\mathcal{A}})$ is called a *Herbrand structure* for F if the following hold:

1. $U_{\mathcal{A}} = D(F)$,

2. For every k-ary function symbol f occuring in F, and for all terms $t_1, t_2, \ldots, t_k \in D(F)$, $f^{\mathcal{A}}(t_1, t_2, \ldots, t_k) = f(t_1, t_2, \ldots, t_k)$.

Example: A Herbrand structure $\mathcal{A} = (U_{\mathcal{A}}, I_{\mathcal{A}})$ for the above example formula F would have the following properties.

$$U_{\mathcal{A}} = D(F) = \{a, f(a), g(a, a), \ldots\}$$

and

$$f^{\mathcal{A}}(a) \;=\; f(a)$$
$$f^{\mathcal{A}}(f(a)) \;=\; f(f(a))$$
$$f^{\mathcal{A}}(g(a, a)) \;=\; f(g(a, a))$$
$$\text{etc.}$$

The choice of $P^{\mathcal{A}}$ is still free. For example, we could define

$(t_1, t_2, t_3) \in P^{\mathcal{A}}$ if and only if $g(t_1, t_2) = g(t_2, f(t_3))$.

This Herbrand structure $\mathcal{A}$ would not be a model for F, because for $t_1 = a$, $t_2 = f(a)$, $t_3 = g(a, a)$ we have that $g(a, f(a)) \neq g(f(a), f(g(a, a)))$.

Exercise 72: Define a Herbrand structure for this example formula which is a model (i.e. modify the definition of $P^{\mathcal{A}}$).

In Herbrand structures the choice of the universe and the interpretation of the function symbols is fixed by definition. What can be chosen freely is the interpretation of the predicate symbols.

At this point, the reader should not proceed before the subtle meaning of clause 2 in the definition of Herbrand structures is understood. There, in a sense, syntax and semantics of terms are synchronized. Terms are interpreted by "themselves". That is, in a Herbrand structure $\mathcal{A}$, for every variable-free term t we have $\mathcal{A}(t) = t$.

Therefore, for Herbrand structures the translation lemma (see Exercise 58) gets the following simplified form

$$\mathcal{A}(F[x/t]) = \mathcal{A}_{[x/t]}(F)$$

that we will use in the following.

We call a Herbrand structure of a formula F a *Herbrand model* for F, simply if it is a model for F.

Theorem

Let F be a closed formula in Skolem form. Then F is satisfiable if and only if F has a Herbrand model.

Proof: It is clear that a formula with a Herbrand model is satisfiable.

Conversely, let $\mathcal{A} = (U_{\mathcal{A}}, I_{\mathcal{A}})$ be an arbitrary model for F. If there is no occurence of a constant symbol in F (this is the special case in the definition of $D(F)$), then we extend $\mathcal{A}$ by the commitment

$$a^{\mathcal{A}} = m,$$

where m is an arbitrary element of $U_{\mathcal{A}}$. This modification of $\mathcal{A}$ does not change the property of being a model. Now, we define a Herbrand structure $\mathcal{B} = (U_{\mathcal{B}}, I_{\mathcal{B}})$ for F. By the definition of Herbrand structures, it remains to define how to interpret the predicate symbols of F as predicates over the Herbrand universe $D(F)$. Let P be any n-ary predicate symbol in F, and let $t_1, t_2, \ldots, t_n \in D(F)$. (Observe that by the above modification of $\mathcal{A}$, $\mathcal{A}(t_1), \ldots, \mathcal{A}(t_n)$ are well defined elements of $U_{\mathcal{A}}$). Now we define

$$(t_1, t_2, \ldots, t_n) \in P^{\mathcal{B}} \text{ if and only if}$$
$$(\mathcal{A}(t_1), \mathcal{A}(t_2), \ldots, \mathcal{A}(t_n)) \in P^{\mathcal{A}}$$

Hence, the definition of $P^{\mathcal{B}}$ "imitates" the definition of $P^{\mathcal{A}}$, by first transforming the arguments $t_1, \ldots, t_n \in D(F) = U_{\mathcal{B}}$ into the universe of $\mathcal{A}$, and then applying $P^{\mathcal{A}}$.

Now we claim that $\mathcal{B}$ is a model for F. Actually, we show a stronger statement: For every closed formula G in prenex form without existential quantifiers that is built up from the same components as F (function symbols and predicate symbols), if $\mathcal{A} \models G$ then $\mathcal{B} \models G$. Then the first claim is the special case $F = G$ in the the second claim. The proof is by induction on the number n of universal quantifiers in G.

In the case $n = 0$, G does not contain a universal quantifier. Then G does not contain a variable. Therefore, immediately from the definition of $\mathcal{B}$, we even get $\mathcal{A}(G) = \mathcal{B}(G)$.

If $n > 0$, then let G be a closed formula in prenex form with n universal quantifiers in the prefix (and no existential quantifiers). Then G has the form $\forall x H$ where H has only $n - 1$ universal quantifiers. We cannot apply the induction hypothesis to H directly because H is not necessarily closed (x could occur free in H). By hypothesis, $\mathcal{A} \models G$, therefore, for all $u \in U_{\mathcal{A}}$, $\mathcal{A}_{[x/u]}(H) = 1$. In particular, for all $u \in U_{\mathcal{A}}$ of the special form $u = \mathcal{A}(t)$ for some $t \in D(G)$, we have $\mathcal{A}_{[x/u]}(H) = 1$. In other words, for all $t \in D(G)$, we have $\mathcal{A}_{[x/\mathcal{A}(t)]}(H) = \mathcal{A}(H[x/t]) = 1$ (by translation lemma). Using the induction hypothesis, $\mathcal{B}(H[x/t]) = 1$ for all $t \in D(G)$. Using the translation lemma again, we have that for all $t \in D(G)$, $\mathcal{B}_{[x/t]}(H) = \mathcal{B}(H[x/t]) = 1$. Hence, $\mathcal{B}(\forall x H) = \mathcal{B}(G) = 1$. ∎

The reader should convince himself that it is relevant for the proof that the formula F is closed, and that F does not contain an existential quantifier.

Corollary (Löwenheim – Skolem)

Every satisfiable formula in predicate logic has a model which is countable (i.e. it has a countable universe).

Proof: Using the methods of Section 2.2, every formula F in predicate logic can be transformed into a s-equivalent closed formula G in Skolem form. Furthermore, these transformations are such that every model of G is also a model of F. Since F is satisfiable, G is satisfiable. Therefore, G possesses a Herbrand model which is, by the above, also a model for F. This Herbrand model has the universe $D(G)$ which is countable.   ∎

Definition (Herbrand expansion)

Let $F = \forall y_1 \forall y_2 \cdots \forall y_n F^*$ be a closed formula in Skolem form. Then $E(F)$, the *Herbrand expansion*, is defined as

$$E(F) = \{F^*[y_1/t_1][y_2/t_2]\cdots[y_n/t_n] \mid t_1, t_2, \ldots, t_n \in D(F)\}$$

That is, the formulas in $E(F)$ are obtained by substituting the terms in $D(F)$ in every possible way for the variables occuring in F^*.

Example: For the above mentioned formula

$$F = \forall x \forall y \forall z P(x, f(y), g(z, x))$$

we obtain the following first elements of $E(F)$,

$$
\begin{array}{rl}
P(a, f(a), g(a, a)) & \text{using} \quad [x/a]\,[y/a]\,[z/a], \\
P(f(a), f(a), g(a, f(a))) & \text{using} \quad [x/f(a)]\,[y/a]\,[z/a], \\
P(a, f(f(a)), g(a, a)) & \text{using} \quad [x/a]\,[y/f(a)]\,[z/a], \\
P(a, f(a), g(f(a), a)) & \text{using} \quad [x/a]\,[y/a]\,[z/f(a)], \\
P(g(a, a), f(a), g(a, g(a, a))) & \text{using} \quad [x/g(a, a)]\,[y/a]\,[z/a], \\
& \text{etc.}
\end{array}
$$

One should observe that the formulas in $E(F)$ can be treated as formulas in *propositional* logic because they do not contain variables. In a sense, instead of $A_1, A_2, \ldots$ another vocabulary is used. To define a structure suitable for all formulas in $E(F)$ it suffices to specify the truth values of the atomic formulas in $E(F)$. The terms (occuring within the atomic formulas) play no role here, and need not be interpreted.

Theorem (Gödel – Herbrand – Skolem)

For each closed formula F in Skolem form, F is satisfiable if and only if the set of formulas $E(F)$ is satisfiable (understood as a set of formulas in propositional logic).

Proof: It suffices to show that F has a Herbrand model if and only if $E(F)$ is satisfiable. Let F have the form $F = \forall y_1 \forall y_2 \ldots \forall y_n F^*$. Then we get:

$\mathcal{A}$ is a Herbrand model for F

> iff for all $t_1, t_2, \ldots, t_n \in D(F)$,
> $\quad \mathcal{A}_{[y_1/t_1][y_2/t_2]\ldots[y_n/t_n]}(F^*) = 1$
> iff for all $t_1, t_2, \ldots, t_n \in D(F)$,
> $\quad \mathcal{A}(F^*[y_1/t_1][y_2/t_2]\ldots[y_n/t_n]) = 1$ (translation lemma)
> iff for all $G \in E(F)$, $\mathcal{A}(G) = 1$
> iff $\mathcal{A}$ is a model for $E(F)$.

∎

This theorem says, in a sense, that predicate logic can be "approximated" by propositional logic. The formula F in predicate logic is associated with $E(F)$, a collection of formulas in propositional logic. The cardinality of $E(F)$ in general is infinite. But by enumerating bigger and bigger finite subsets of $E(F)$, F can be approximated (or better: the question of F's satisfiablity can be approximated).

The issue of finite subsets of infinite sets of formulas in propositional logic brings up the possibility of applying the compactness theorem proved in Section 1.4. This is done in the following theorem.

Herbrand's Theorem

A closed formula in Skolem form is unsatisfiable if and only if there is a finite subset of $E(F)$ which is unsatisfiable (in the sense of propositional logic).

Proof: A direct combination of the previous theorem and the compactness theorem for propositional logic (Section 1.4).   ∎

Based on Herbrand's theorem, so-called semi-decision procedures for predicate logic can be formulated. A semi-decision procedure for a problem (as introduced in Section 2.3) is understood as a program that stops exactly for those instances after finitely many steps for which the question has to be answered "yes".

The following is a semi-decision procedure for the unsatisfiability problem. Its correctness follows immediately from Herbrand's theorem. For the presentation of the program, we think of the formulas in $E(F)$ as being enumerated:

$$E(F) = \{F_1, F_2, \ldots, F_n, \ldots\}$$

Because Gilmore was one of the first to implement a simple semi-decision procedure for predicate logic based directly on Herbrand's theorem, we call the following procedure *Gilmore's procedure.*

Gilmore's Procedure

Instance: A closed formula F in Skolem form (every formula in predicate logic can be transformed into a s-equivalent formula of this kind, cf. Section 2.2).

$n := 0$;

repeat $n := n + 1$;

until $(F_1 \wedge F_2 \wedge \cdots \wedge F_n)$ is unsatisfiable (this can be tested with the tools of propositional logic, e.g. using truth tables) ;

output "unsatisfiable" and halt;

This program has the property that it stops after finitely many steps on every unsatisfiable formula as input, and for satisfiable formulas, it does not stop. This is exactly what is needed for semi-decidability: on the "yes-instances" the program stops, but not on the "no-instances." By testing $\neg F$ for unsatisfiability, we obtain a semi-decision procedure for validity. Therefore, we can summarize:

Theorem

1. The unsatisfiability problem for formulas in predicate logic is semi-decidable.

2. The validity problem for formulas in predicate logic is semi-decidable.

Exercise 73: Show that the notion of semi-decidability introduced here is equivalent to the notion of *recursive enumerability*. A set M (the set of yes-instances of a given problem) is recursively enumerable if $M = \emptyset$ or if there is a total function f which is effectively computable such that $M = \{f(1), f(2), f(3), \ldots\}$. In the example above, the set M would be the set of unsatisfiable formulas in predicate logic.

Exercise 74: Show that a problem is decidable if and only if it is recursively enumerable (see last exercise) in such a way that the enumerating function is nondecreasing: $f(n) \leq f(n+1)$ for all n.

Exercise 75: Show that the PCP (see Section 2.3) is semi-decidable.

Combining the unsatisfiability test and the validity test, we can obtain a procedure which stops on the unsatisfiable formulas and on the valid formulas (with respective output "unsatisfiable" or "valid"). Furthermore, one could patch a third procedure which on a given input formula F systematically searches for models of finite cardinality $n = 1, 2, 3, \ldots$. Combined this gives a procedure that stops after finitely many steps when applied to formulas in the marked areas – with corresponding output.

all formulas in predicate logic

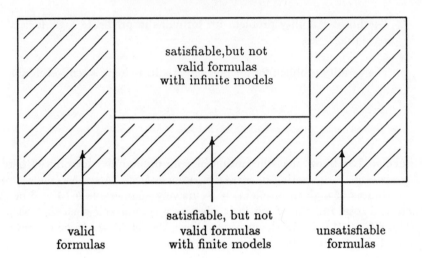

valid
formulas

satisfiable, but not
valid formulas
with finite models

unsatisfiable
formulas

The white area in the diagram could be reduced further somewhat (e.g. for formulas of certain syntactical properties), but it can never be eliminated or become finite. This would be in contradiction to the undecidability result proved in Section 2.3.

2.5 Resolution

The tests for unsatisfiability on the finite subsets of $E(F)$ which have to be performed in Gilmore's procedure could as well be implemented by resolution. For this, we have to presuppose that the matrix of F is in CNF. (This can always be achieved, see Sections 1.2 and 2.2). All formulas in $E(F)$ result from certain substitutions for the variables in F^*. Therefore, all formulas in $E(F)$ are in CNF provided that F^* is in CNF.

If a formula G results from certain substitutions from a formula F, then G is called an *instance* of F. Substitutions which make a formula variable-free (like in the definition of $E(F)$) are called *ground substitutions*, and the result of applying a ground substitution to a formula is a *ground instance* of that formula. Thus, the following modification of Gilmore's procedure is called the *ground resolution procedure*. Its correctness follows immediately from the correctness of Gilmore's procedure.

In the following, we assume again that $E(F)$ is enumerated as $F_1, F_2, \ldots$. (Remember that $Res^*(\ )$ was defined in Section 1.5.)

Ground Resolution Procedure

Instance: a closed formula F in Skolem form with its matrix F^* in **CNF**

$i := 0;$
$M := \emptyset;$
repeat
$\quad i := i + 1;$
$\quad M := M \cup \{F_i\};$
$\quad M := Res^*(M);$
until $\square \in M;$
Output "unsatisfiable" and halt;

Combining Herbrand's Theorem and the resolution theorem of propositional logic, we obtain the following theorem.

Theorem

Using as input any closed formula F in Skolem form where the matrix F^* is in **CNF**, the ground resolution procedure stops after a finite number of steps if and only if F is unsatisfiable.

Similar to the resolution algorithm in propositional logic, it is usually the case that more elements are generated in M than are really needed for the "demonstration" of unsatisfiability of the input formula F (and in the case of a satisfiable formula as input, in general infinitely many elements are generated in M). Relevant for the demonstration of unsatisfiability are such formulas occuring in the resolution graph of the first finite subset of

$E(F)$ which is unsatisfiable. For such a "demonstration" of unsatisfiability of F, it suffices to specify certain ground substitutions for F^* first (leading to certain elements of $E(F)$) and then to present a resolution proof based on these ground instances.

Example: Consider the following unsatisfiable formula

$$F = \forall x(P(x) \wedge \neg P(f(x))).$$

Here we have,

$$F^* = (P(x) \wedge \neg P(f(x))),$$

which is written in clause form,

$$F^* = \{\{P(x)\}, \{\neg P(f(x))\}\}.$$

Furthermore,

$$E(F) = \{(P(a) \wedge \neg P(f(a))), (P(f(a)) \wedge \neg P(f(f(a)))), \ldots\}.$$

Already the first two ground substitutions $[x/a]$ and $[x/f(a)]$ lead to a finite unsatisfiable clause set. This corresponds to the first two formulas in $E(F)$, which form four clauses as listed below.

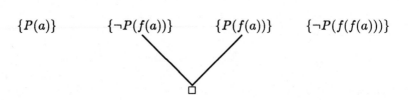

In this example, already two clauses are generated (as part of the first and second formulas in $E(F)$) which are not needed for the resolution refutation. Therefore, we conclude that it suffices to consider ground substitutions that are applied *individually* to the clauses of the original formula F^*.

We express this situation by the following diagram where vectors are used to express (ground) substitutions.

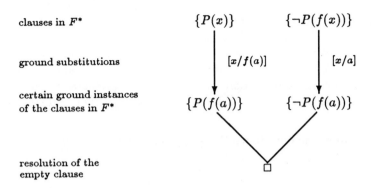

clauses in F^*

ground substitutions

certain ground instances
of the clauses in F^*

resolution of the
empty clause

Let us consider a more complex example. Let

$$F = \forall x \forall y ((\neg P(x) \vee \neg P(f(a)) \vee Q(y)) \wedge P(y) \wedge (\neg P(g(b,x)) \vee \neg Q(b))).$$

Then we obtain the following clause representation of F^*,

$$F^* = \{\{\neg P(x), \neg P(f(a)), Q(y)\}, \{P(y)\}, \{\neg P(g(b,x), \neg Q(b)\}\}.$$

This formula F is unsatisfiable. A proof for the unsatisfiability of F is given
by the following diagram.

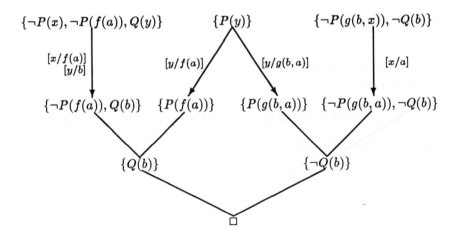

Again, vectors denote ground substitutions. In this example two new aspects occur. First, it might be necessary to use the same clause in F^*

to derive several ground instances from it to enable the resolution refutation. (This is the case for the clause $\{P(y)\}$). Second, from an n-element clause an m-element clause can be obtained after the ground resolution step ($m \leq n$). We get $m < n$ if certain literals in the original clause become identical after the substitution, and by the set representation melt into a single element. (This is the case for the clause $\{\neg P(x), \neg P(f(a)), Q(y)\}$ and the substitution $[x/f(a)][y/b]$).

We summarize our observations in the following theorem.

Theorem (ground resolution theorem)

A closed formula F in Skolem form $F = \forall y_1 \forall y_2 \ldots \forall y_k F^*$ with its matrix F^* in **CNF** is unsatisfiable if and only if there exists a finite sequence of clauses $C_1, C_2, \ldots, C_n$ with the properties

C_n is the empty clause, and for $i = 1, \ldots, n$,

either C_i is a ground instance of some clause $C \in F^*$,
i.e. C_i has the form $C_i = C[y_1/t_1][y_2/t_2] \cdots [y_k/t_k]$
where $t_1, t_2, \ldots, t_k \in D(F)$,
or C_i is a resolvent (in the sense of propositional logic)
of two clauses C_a and C_b with $a, b < i$.

Exercise 76: Formalize the following statements 1 and 2 as formulas in predicate logic

(a) The professor is happy if all his students like logic.

(b) The professor is happy if he has no students.

and show, by ground resolution, that (b) is a consequence of (a).

The algorithmic selection of ground instances of F^* which allows one to perform a resolution refutation afterwards, does not seem to be programmable in a "controlled" way, just by exhaustive search. The problem is that certain decisions for substitutions have to be done in a "lookahead" manner to enable resolution steps further "down" in the resolution graph. This difficulty suggests a modification, namely not to perform all substitutions in the beginning, but rather in a successive "on demand" manner. Here, the demand comes from the resolution step that directly follows. But this requires that resolution steps be performed with clauses in predicate logic.

Now we introduce the predicate logic version of resolution which was invented by J. A. Robinson. The new idea is to resolve clauses in predicate logic to clauses in predicate logic where each resolution step is accompanied by a substitution. These substitutions are performed in a guarded manner. For example, in the case of the two clauses $\{P(x), \neg Q(g(x))\}$ and $\{\neg P(f(y))\}$, it suffices to use the substitution $[x/f(y)]$ to obtain the resolvent $\{\neg Q(g(f(y)))\}$. There is no need at this point to substitute anything for the variable y.

Central for the following investigations is the search for a substitution which *unifies* two or more literals, i.e., makes them identical. In the above example, $[x/f(y)]$ unifies the two literals $P(x)$ and $P(f(y))$. The substitution $[x/f(a)][y/a]$ would also be a unifier but does not satisfy the definition of a *most general unifier*. In a sense (defined formally below), this substitution makes more substitutions than necessary.

Definition (unifier, most general unifier)

A substitution *sub* is a *unifier* for a (finite) set of literals $L = \{L_1, L_2, \ldots, L_k\}$, if $L_1 sub = L_2 sub = \ldots = L_k sub$.

That is, by applying *sub* to every literal in the set L, one and only one literal is obtained. If $L sub$ expresses the set obtained by applying *sub* to every literal in the set L, then this situation can be formally expressed by $|L sub| = 1$. If a substitution *sub* exists with the property that $|L sub| = 1$, then we say L is *unifiable*.

A unifier *sub* for some literal set L is called a *most general unifier* if for every unifier *sub'* there is a substitution *s* such that $sub' = sub\, s$. (Here, the equality $sub' = sub\, s$ means that for every formula F, $F sub' = F sub\, s$).

The following diagram describes the situation.

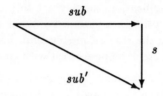

Unification Theorem (Robinson)

Every unifiable set of literals has a most general unifier.

Proof: We prove this theorem *constructively* in the sense that an algorithm is presented, which takes as input a set of literals **L**, and terminates after finitely many steps either with the output "unifiable" or "non-unifiable". Further, in the case of unifiability, it also outputs a most general unifier. A proof of correctness of such an algorithm is also a proof for the assertion of the theorem. Now we describe this algorithm.

Unification Algorithm

Instance: A non-empty set of literals **L**.

$sub := [\ ]$; (this is the empty substitution)
while $|\mathbf{L}sub| > 1$ **do**
    begin
        Scan the literals in $\mathbf{L}sub$ from left to right, until the first
        position is found where in at least two literals (say, L_1 and
        L_2) the corresponding symbols are different ;
        if none of these symbols is a variable **then**
            output "non-unifiable" and halt
        else
            begin
                Let x be the variable, and let t be a term that is
                different from x and which starts at this position
                in another literal (this can also be a variable) ;
                if x occurs in t **then**
                    output "non-unifiable" and halt ;
                else $sub := sub[x/t]$;
                    (this means the composition of the sub-
                    stitutions sub and $[x/t]$)

> **end**
> **end;**

output sub as a most general unifier of **L** ;

For the correctness of this algorithm, we first observe that it always terminates, because in each application of the **while** loop another variable x is substituted by a term t (in which x does not occur). Therefore the number of different variables occuring in **L**sub decreases by 1 in each step. Hence there are at most as many applications of the **while** loop as there are different variables in **L** in the beginning.

If the algorithm terminates successfully and leaves the **while** loop, then the output sub must necessarily be a unifier for **L**, because the **while** loop is only left if $|\mathbf{L}sub| = 1$. Since we have shown that the algorithm always terminates, in case of a non-unifiable clause set **L** as input, the algorithm necessarily has to stop inside the **while** loop and outputs correctly "non-unifiable".

It remains to show that in case of a unifiable set of literals **L** as input, indeed a *most general* unifier is produced. Let sub_i be the substitution which is obtained after the ith application of the **while** loop. Then we have $sub_0 = [\ ]$. We show by induction on i that in case of a unifiable set of literals **L**, for every unifier sub' of **L**, there is a substitution s_i such that $sub' = sub_i\, s_i$, and that the **while** loop is either successfully left in the i-th step, or both **else** branches in the **while** loop are entered (in which case the **while** loop can be executed for another time.) From this, it follows that the **while** loop is finally left successfully, say after the n-th loop, and the output sub_n satisfies the definition of a most general unifier.

If $i = 0$, then we let $s_0 = sub'$. Then we have $sub' = s_0 = [\ ]\, s_0 = sub_0\, s_0$.

For $i > 0$, let s_{i-1} be the substitution which exists by induction hypothesis with $sub' = sub_{i-1}s_{i-1}$. Now, either $|\mathbf{L}sub_{i-1}| = 1$, and the **while** loop is left successfully, or $|\mathbf{L}sub_{i-1}| > 1$ and the **while** loop is entered for the i-th time. By the fact that $|\mathbf{L}sub_{i-1}| > 1$ and since sub_{i-1} can be extended to a unifier of **L** by applying s_{i-1}, there must exist some variable x and a different term t (at a position where two literals L_1 and L_2 in $\mathbf{L}sub_{i-1}$ differ) so that x does not occur in t. Therefore both **else** branches will be entered. Hence, s_{i-1} unifies x and t, i.e. $xs_{i-1} = ts_{i-1}$. Furthermore, sub_i is then set in the i-th loop to $sub_{i-1}[x/t]$. Now we modify the substitution s_{i-1} so that we take out any replacement for the variable x (but all other substitions in s_{i-1} remain.) Let the result of this restriction be s_i. We claim that s_i has the desired properties. We have

$$
\begin{aligned}
sub_i s_i \; &= \; sub_{i-1}[x/t]s_i \\
&= \; sub_{i-1}s_i[x/ts_i] && \text{because } x \text{ is not substituted in } s_i \\
&= \; sub_{i-1}s_i[x/ts_{i-1}] && \text{because } x \text{ does not occur in } t \\
&= \; sub_{i-1}s_{i-1} && \text{because } xs_{i-1} = ts_{i-1} \\
& && \text{and the definition of } s_i \\
&= \; sub' && \text{by induction hypothesis}
\end{aligned}
$$

This completes the proof of the unification theorem.   ∎

Example: We want to apply the unification algorithm to the set of literals

$$\mathbf{L} = \{\neg P(f(z, g(a, y)), h(z)), \neg P(f(f(u, v), w), h(f(a, b)))\}.$$

Then we obtain in the first step

$$
\begin{aligned}
&\neg P(f(z, g(a, y)), h(z)) \\
&\neg P(f(f(u, v), w), h(f(a, b))) \\
&\qquad\quad \uparrow
\end{aligned}
$$

which results in the substitution $sub = [z/f(u, v)]$. In the second step, after applying sub, we obtain:

$$
\begin{aligned}
&\neg P(f(f(u, v), g(a, y)), h(f(u, v))) \\
&\neg P(f(f(u, v), w), h(f(a, b))) \\
&\qquad\qquad\qquad \uparrow
\end{aligned}
$$

Therefore, the substitution is extended by $[w/g(a, y)]$. Next, we obtain

$$
\begin{aligned}
&\neg P(f(f(u, v), g(a, y)), h(f(u, v))) \\
&\neg P(f(f(u, v), g(a, y)), h(f(a, b))) \\
&\qquad\qquad\qquad\qquad \uparrow
\end{aligned}
$$

Now sub is extended by $[u/a]$. In the fourth step

$$
\begin{aligned}
&\neg P(f(f(a, v), g(a, y)), h(f(a, v))) \\
&\neg P(f(f(a, v), g(a, y)), h(f(a, b))) \\
&\qquad\qquad\qquad\qquad \uparrow
\end{aligned}
$$

we obtain the final substitution $sub = [z/f(u, v)][w/g(a, y)][u/a][v/b]$. This is a most general unifier for $\mathbf{L}$, and we have

$$\mathbf{L}sub = \{\neg P(f(f(a, b)), g(a, y)), h(f(a, b)))\}.$$

Observe that *sub* is not a ground substitution for **L** since the variable y still occurs in **L***sub*.

In some situations, it is desirable to write down substitutions in a "disentangled" way so that all partial substitutions can be applied in any order – or in parallel – without changing the result. A disentangled version of the above substitution *sub* is

$$sub = [z/f(a, b)][w/g(a, y)][u/a][v/b].$$

Exercise 77: Show how for two disentangled substitions *sub* and *sub'*, their concatenation *sub sub'* can be disentangled again.

Exercise 78: Apply the unification algorithm to the set of literals

$$\mathbf{L} = \{P(x, y), P(f(a), g(x)), P(f(z), g(f(z)))\}.$$

Exercise 79: Show that the unification algorithm (implemented in a straightforward way) can have exponential running time.

Hint: Consider the example

$$\mathbf{L} = \{P(x_1, x_2, \ldots, x_n), P(f(x_0, x_0), f(x_1, x_1), \ldots, f(x_{n-1}, x_{n-1}))\}.$$

Think of a data structure for literals and sets of literals which allows a more efficient implementation of the unification algorithm.

Exercise 80: In some implementations of the unification algorithm (e.g. in interpreters for the programming language PROLOG), by efficiency reasons, the test "does x occur in t" is left out (the *occurence check*).

Give an example of a 2-element set $\mathbf{L} = \{L_1, L_2\}$ which is not unifiable. Let L_1 and L_2 have no variables in common, and (still!) a unification algorithm without occurence check gets into an infinite loop (or erroneously outputs that **L** is unifiable – depending on the implementation).

Using the unification principle, we are now in a situation to formulate the resolution principle for predicate logic.

Definition (resolution in predicate logic)

Let C_1, C_2 and R be clauses (in predicate logic). Then R is called a *resolvent* of C_1, C_2 if the following holds.

1. There exist certain substitutions s_1 and s_2 which are variable renamings so that $C_1 s_1$ and $C_2 s_2$ do not contain the same variable.

2. There is a set of literals $L_1, \ldots, L_m \in C_1 s_1$ $(m \geq 1)$ and $L'_1, \ldots, L'_n \in C_2 s_2$ $(n \geq 1)$, such that $\mathbf{L} = \{\overline{L_1}, \overline{L_2}, \ldots, \overline{L_m}, L'_1, L'_2, \ldots, L'_n\}$ is unifiable. Let *sub* be a most general unifier for $\mathbf{L}$.

3. R has the form

$$R = ((C_1 s_1 - \{L_1, \ldots, L_m\}) \cup (C_2 s_2 - \{L'_1, \ldots, L'_n\})) sub .$$

We express the situation described by the definition by the following diagram.

For better legibility, the literals $L_1, \ldots, L_m, L'_1, \ldots, L'_n$ can be underlined, and the substitutions used can be noted beside the diagram.

Example:

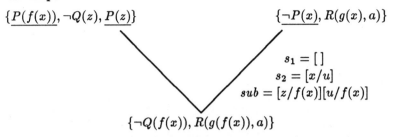

$\{\underline{P(f(x))}, \neg Q(z), \underline{P(z)}\}$ $\{\underline{\neg P(x)}, R(g(x), a)\}$

$$s_1 = [\,]$$
$$s_2 = [x/u]$$
$$sub = [z/f(x)][u/f(x)]$$

$\{\neg Q(f(x)), R(g(f(x)), a)\}$

Remark: The resolution calculus in propositional logic can be understood as a special case of resolution in predicate logic where $s_1 = s_2 = sub = [\,]$ and $m = n = 1$. Therefore, we adopt the notation introduced for the resolution in the propositional calculus, and extend the notion $Res(F)$ also for clause sets in predicate calculus:

$$Res(F) = F \cup \{R \mid R \text{ is a resolvent}$$
$$\text{of two clauses } C_1, C_2 \in F\},$$
$$Res^0(F) = F,$$
$$Res^{n+1}(F) = Res(Res^n(F)) \text{ for } n \geq 0,$$
$$\text{and}$$
$$Res^*(F) = \bigcup_{n \geq 0} Res^n(F).$$

As in propositional logic, it is clear that $\square \in Res^*(F)$ if and only if there is a sequence $C_1, C_2, \ldots, C_n$ of clauses such that $C_n = \square$, and for $i = 1, 2, \ldots, n$, C_i is either element of F or C_i is resolvent of two clauses C_a and C_b with $a, b < i$.

Exercise 81: Find all resolvents of the following two clauses C_1 and C_2.

$$C_1 = \{\neg P(x, y), \neg P(f(a), g(u, b)), Q(x, u)\}$$
$$C_2 = \{P(f(x), g(a, b)), \neg Q(f(a), b), \neg Q(a, b)\}$$

As preparation for the proof of the resolution theorem, we show how resolutions in propositional calculus (for ground instances of clauses in predicate logic) can be "lifted" to certain resolutions in predicate logic. This

"Lifting-Lemma" allows us to transform a resolution refutation on clauses in propositional logic to a resolution refutation on clauses in predicate logic.

Lifting-Lemma

Let C_1, C_2 be two clauses in predicate logic und let C_1', C_2' be two arbitrary ground instances thereof which are resolvable (in the sense of propositional logic). Let R' be a resolvent of C_1', C_2'. Then there exists a clause R which is resolvent of C_1, C_2 (in the sense of predicate logic) so that R' is a ground instance of R.

The following two pictures demonstrate the situation.

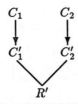

Assumption of the Lifting-Lemma

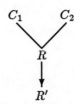

Conclusion of the Lifting-Lemma

Proof: First, let s_1 and s_2 be variable renamings such that $C_1 s_1$ and $C_2 s_2$ do not have a variable in common. Since C_1' and C_2' are ground instances of C_1 and C_2, they are also ground instances of $C_1 s_1$ and $C_2 s_2$. Let sub_1, sub_2 be ground substitutions such that $C_1' = C_1 s_1 sub_1$ and $C_2' = C_2 s_2 sub_2$. Since there is no variable which is replaced in both substitutions sub_1 and sub_2, we let $sub = sub_1 sub_2$, and we get $C_1' = C_1 s_1 sub$ and $C_2' = C_2 s_2 sub$. By assumption, C_1' and C_2' have some resolvent R' (in propositional logic). Therefore, there must be a literal $L \in C_1'$ such that $\overline{L} \in C_2'$ and $R' = (C_1' - \{L\}) \cup (C_2' - \{\overline{L}\})$. The literal L results from one or more literals in $C_1 s_1$ by the ground substitution sub. The same holds for $\overline{L}$ and $C_2 s_2$. Hence there are literals $L_1, \ldots, L_m \in C_1 s_1 (m \geq 1)$ and $L_1', \ldots, L_n' \in C_2 s_2 (n \geq 1)$, such that $L = L_1 sub = \ldots = L_m sub$ and $\overline{L} = L_1' sub = \ldots = L_n' sub$. Therefore, $C_1 s_1$, $C_2 s_2$ (and also C_1, C_2) are resolvable, because sub is a unifier for the set of literals

$$\mathbf{L} = \{L_1, \ldots, L_m, \overline{L_1'}, \ldots, \overline{L_n'}\}.$$

Let sub_0 be a most general unifier for $\mathbf{L}$ provided by the unification algorithm. Then,

$$R = ((C_1 s_1 - \{L_1, \ldots, L_m\}) \cup (C_2 s_2 - \{L_1', \ldots, L_n'\})) sub_0$$

is a (predicate logic) resolvent of $C_1 s_1$, $C_2 s_2$ (and also of C_1, C_2). Since sub_0 is a most general unifier and sub is a unifier of $\mathbf{L}$, there exists a substitution s such that $sub_0 s = sub$. Therefore, we get

$$
\begin{aligned}
R' &= (C_1' - \{L\}) \cup (C_2' - \{\overline{L}\}) \\
&= (C_1 s_1 sub - \{L\}) \cup (C_2 s_2 sub - \{\overline{L}\}) \\
&= ((C_1 s_1 - \{L_1, \ldots, L_m\}) \cup (C_2 s_2 - \{L_1', \ldots, L_n'\})) sub \\
&= ((C_1 s_1 - \{L_1, \ldots, L_m\}) \cup (C_2 s_2 - \{L_1', \ldots, L_n'\})) sub_0 s \\
&= R s
\end{aligned}
$$

This shows that R' is a ground instance of R (via the substitution s). ∎

Exercise 82: Consider the following ground resolution.

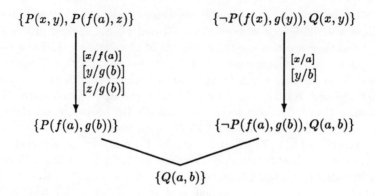

Follow the proof of the Lifting Lemma, and find out which (predicate logic) resolution step is constructed from this.

Resolution Theorem (of predicate logic)

Let F be a closed formula in Skolem form with its matrix F^* in **CNF**. Then, F is unsatisfiable if and only if $\square \in Res^*(F^*)$.

Proof: (Correctness) First we show that $\square \in Res^*(F^*)$ implies that F is unsatisfiable. For a formula H with the free variables $x_1, x_2, \ldots, x_n$ let $\forall H$ denote its *universal closure*. This is the formula $\forall H = \forall x_1 \forall x_2 \ldots \forall x_n H$. Note that $F \equiv \bigwedge_{C \in F^*} \forall C$. Now we show that for every resolvent R of two clauses C_1, C_2, $\forall R$ is a consequence of $\forall C_1 \wedge \forall C_2$. Then, it follows that the empty clause is a consequence of F, and therefore, F is unsatisfiable.

Let $\mathcal{A}$ be a structure such that $\mathcal{A}(\forall C_1) = \mathcal{A}(\forall C_2) = 1$. Let the resolvent R have the form

$$
\begin{aligned}
R &= ((C_1 s_1 - \{L_1, \ldots, L_m\}) \cup (C_2 s_2 - \{L_1', \ldots, L_n'\})) sub \\
  &= (C_1 s_1 sub - \{L\}) \cup (C_2 s_2 sub - \{\overline{L}\}),
\end{aligned}
$$

where sub is a most general unifier of $\mathbf{L} = \{L_1, \ldots, L_m, \overline{L_1'}, \ldots, \overline{l_n'}\}$, and $L = L_1 sub = \ldots = L_m sub = \overline{L_1'} sub = \ldots = \overline{L_n'} sub$. Assume that $\mathcal{A}(\forall R) = 0$. Then there exists a structure $\mathcal{A}'$ with $\mathcal{A}'(R) = 0$, where $\mathcal{A}'$ is the same as $\mathcal{A}$, but additionally has suitable interpretations for the variables that occur. Then we have $\mathcal{A}'(C_1 s_1 sub - \{L\}) = 0$ and $\mathcal{A}'(C_2 s_2 sub - \{\overline{L}\}) = 0$. Because of $1 = \mathcal{A}'(C_1 s_1 sub) = \mathcal{A}'(C_2 s_2 sub)$, it follows that $\mathcal{A}'(L) = \mathcal{A}'(\overline{L}) = 1$. This is a contradiction which shows that $\mathcal{A}(\forall R) = 1$.

(Completeness) Suppose that F is unsatisfiable. Using the ground resolution theorem, there is a sequence of clauses $(C'_1, C'_2, \ldots, C'_n)$ such that $C'_n = \square$, and for $i = 1, 2, \ldots, n$, C'_i either is ground instance of some clause in F^* or C'_i is a (propositional logic) resolvent of two clauses C'_a and C'_b with $a, b < i$. For $i = 1, 2, \ldots, n$ we now construct a sequence C_i of predicate logic clauses where $C_n = \square$ which demonstrates that $\square \in Res^*(F)$. If C'_i is a ground instance of some clause $C \in F^*$, then we choose $C_i = C$. If C'_i is resolvent of two clauses C'_a and C'_b with $a, b < i$, then we have already determined the clauses C_a and C_b such that C'_a and C'_b are ground instances thereof. By the Lifting Lemma, we can find a clause C_i which is resolvent of C_a and C_b, and such that C'_i is ground instance of C_i. The sequence $(C_1, C_2, \ldots, C_n)$ that is obtained shows that $\square \in Res^*(F)$. ∎

Example: The clause set

$$F = \{\{\neg P(x), Q(x), R(x, f(x))\}, \{\neg P(x), Q(x), S(f(x))\}, \{T(a)\},$$
$$\{P(a)\}, \{\neg R(a, z), T(z)\}, \{\neg T(x), \neg Q(x)\}, \{\neg T(y), \neg S(y)\}\}$$

is unsatisfiable. A deduction of the empty clause is given by

(1) $\{T(a)\}$ .... clause in F

(2) $\{\neg T(x), \neg Q(x)\}$ .... clause in F

(3) $\{\neg Q(a)\}$ .... resolvent of (1) and (2)

(4) $\{\neg P(x), Q(x), S(f(x))\}$ .... clause in F

(5) $\{P(a)\}$ .... clause in F

(6) $\{Q(a), S(f(a))\}$ .... resolvent of (4) and (5)

(7) $\{S(f(a))\}$ .... resolvent of (3) and (6)

(8) $\{\neg P(x), Q(x), R(x, f(x))\}$ .... clause in F

(9) $\{Q(a), R(a, f(a))\}$ .... resolvent of (5) and (8)

(10) $\{R(a, f(a))\}$ .... resolvent of (3) and (9)

(11) $\{\neg R(a, z), T(z)\}$ .... clause in F

(12) $\{T(f(a))\}$ .... resolvent of (10) and (11)

(13) $\{\neg T(y), \neg S(y)\}$ .... clause in F

(14) $\{\neg S(f(a))\}$ resolvent of (12) and (13)

(15) □   resolvent of (7) and (14)

Exercise 83: For finite clause sets F in propositional logic, $Res^*(F)$ is always a finite set. Show that there are finite clause sets F in predicate logic such that for all n,

$$Res^n(F) \neq Res^*(F).$$

Example: To demonstrate the use of the resolution calculus for automated theorem proving, we consider the following example from group theory. Let ∘ be the group operation. By $P(x, y, z)$ we express that $x \circ y = z$. Then the axioms of group theory can be expressed by the following formulas.

(1) $\forall x \forall y \exists z P(x, y, z)$
    (closure under ∘)

(2) $\forall u \forall v \forall w \forall x \forall y \forall z ((P(x, y, u) \wedge P(y, z, v)) \rightarrow (P(x, v, w) \leftrightarrow P(u, z, w)))$
    (associativity)

(3) $\exists x (\forall y P(x, y, y) \wedge \forall y \exists z P(z, y, x))$
    (existence of a left-neutral element
    and existence of left-inverses)

Now we want to prove that the existence of right-inverses follows from (1), (2), and (3). This is expressed by the following formula (4).

(4) $\exists x (\forall y P(x, y, y) \wedge \forall y \exists z P(y, z, x))$

Converting $(1) \wedge (2) \wedge (3) \wedge \neg(4)$ into clause form gives

(a) $\{P(x, y, m(x, y))\}$

(b) $\{\neg P(x, y, u), \neg P(y, z, v), \neg P(x, v, w), P(u, z, w)\}$

(c) $\{\neg P(x, y, u), \neg P(y, z, v), \neg P(u, z, w), P(x, v, w)\}$

(d) $\{P(e, y, y)\}$

(e) $\{P(i(y), y, e)\}$

(f) $\{\neg P(x, j(x), j(x)), \neg P(k(x), z, x)\}$

Here, m (2-ary), e (0-ary), i (1-ary), and k (1-ary) are newly introduced Skolem functions. A resolution refutation from (a)–(f), and therefore, a proof of unsatisfiablity is given by the following diagram (which happens to be a linear chain).

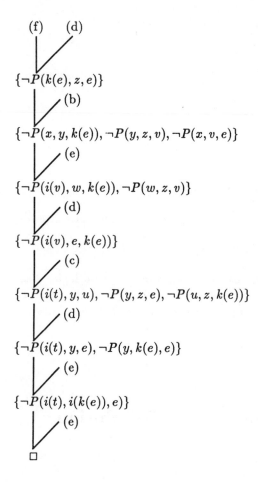

Exercise 84: Show that the following are consequences of the above mentioned axioms of group theory.

(a) There exists a right-neutral element.

(b) If G is an Abelian group, then for all x, y in G, $x \circ y \circ x^{-1} = y$.

Exercise 85: Express the following facts by formulas in predicate logic.

(a) Every dragon is happy if all its children can fly.

(b) Green dragons can fly.

(c) A dragon is green if it is a child of at least one green dragon.

Show by resolution that from (a),(b) and (c) follows that all green dragons are happy.

Exercise 86: Given are the following facts.

(a) Every barber shaves all persons who do not shave themselves.

(b) No barber shaves any person who shaves himself.

Formalize (a) and (b) as formulas in predicate logic. Use $B(x)$ for "x is barber", and $S(x, y)$ for "x shaves y". Convert into clausal form and show by resolution that (c) is a consequence of (a) and (b).

(c) There are no barbers.

2.6   Refinements of Resolution

Although the predicate logic version of resolution constitutes a great improvement as compared to the straightforward ground resolution procedure, there is a tremendous combinatorial explosion with which one has to deal. The problem is that, in general, there are many possiblities to find two resolvable clauses for producing new resolvents. Among this huge number of possible resolution steps, only a few might lead to the derivation of the empty clause (in case the clause set is unsatisfiable). Additionally, while the resolution process proceeds, the number of clauses (and their lengths) increases further, which causes still more choices to be tried.

We now present some possibilities of improving the efficiency of the general resolution algorithm. We call these *refinements* of resolution. We distinguish between *strategies* and *restrictions*.

Strategies are just heuristic rules which prescribe the (deterministic) order through which the (nondeterministic) search space has to be explored. Hence, the *size* of the search space is not affected by a strategy. But for a clever strategy, there is some hope that only a small portion of the space has to be searched until a solution (a derivation of the empty clause) is found. In the worst case, the entire space has to be searched.

An example is the *unit preference strategy* where, whenever possible, resolution steps are performed when one of the parent clauses is a *unit*, i.e. consists of one literal only.

These strategies seem to work quite well in the examples studied, but there is little theoretical work which can be reported here. We just mention that such strategies can be combined with the resolution restrictions which will be discussed next.

The resolution restrictions however simply forbid certain resolution steps if the clauses involved do not have a certain syntactic form, depending on the type of restriction. Therefore, the number of possible choices for the next resolution step is smaller as compared to the general case. Of course, the question to be investigated is whether such restrictions go "too far", so that the calculus loses the completeness property. (This would be the case if there is an unsatisfiable clause set such that the empty clause is not derivable under the respective restriction).

We now present the different resolution restrictions that we will study in the following.

The *P-restriction* (or just *P-resolution*) requires that at least one of the parent clauses has to be positive, i.e., consists of positive literals only. Analogously, the *N-restriction* (*N-resolution*) requires that at least one parent clause is negative. We will later show that P-resolution as well as N-resolution are complete.

The empty clause is *linearly resolvable* from a clause set F, *based on* a clause $C \in F$, if there is a sequence of clauses $(C_0, C_1, \ldots, C_n)$ such that $C_0 = C$, $C_n = \Box$, and for $i = 1, 2, \ldots, n$,

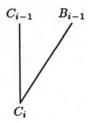

where the clause B_{i-1} (the so-called *side clause*) is either an element of F (i.e. an *input clause*) or $B_{i-1} = C_j$ for some $j < i$.

We will show in this section that linear resolution is complete, that is, for every unsatisfiable clause set F there is a clause $C \in F$ (called the *base clause*) such that the empty clause is linearly resolvable from F based on C.

Example: Consider the unsatisfiable clause set

$$F = \{\{A, B\}, \{A, \neg B\}, \{\neg A, B\}, \{\neg A, \neg B\}\}.$$

The usual resolution refutation is given by the following diagram and takes 3 resolution steps.

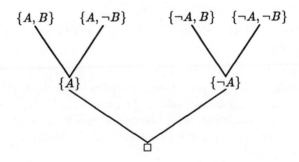

A linear resolution of the empty clause from F, based on $\{A, B\}$, is given by the following diagram (this is also an example for a P-resolution).

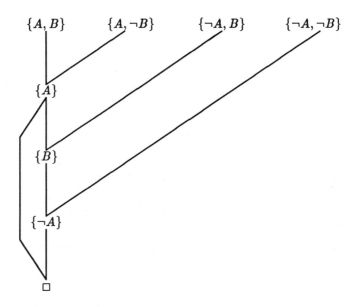

Observe that this resolution refutation consists of 4 resolution steps. This suggests that the price to be paid for the restriction in the number of nondeterministic choices is an increase in the proof length. This effect is not yet theoretically explored (cf. Exercise 87).

For the *set-of-support restriction* of resolution one needs to know (e.g. from the context) a subset T of the clause set F such that $F - T$ is satisfiable. A resolution deduction of the empty clause from F, relative to the set-of-support T, has to satisfy the requirement that it is never the case that two clauses from $F - T$ are being resolved. This restriction can bring an advantage if T is relatively small (e.g. $|T| = 1$) and therefore, $F - T$ is relatively big. Many potential resolution steps (between clauses in $F - T$) can be avoided this way. A typical example is to test whether a given formula G follows from the "data base" $\{F_1, F_2, \ldots, F_n\}$. We know that this is the case if and only if the set $\{F_1, F_2, \ldots, F_n, \neg G\}$ is unsatisfiable (Exercise 3). If it is known from the context that $\{F_1, F_2, \ldots, F_n\}$ is satisfiable, i.e., the data base (or axiom system, if you prefer) is consistent − or if this consistency is just assumed, then one can choose as set-of-support $T = \{G_1, \ldots, G_k\}$ where $\{G_1, \ldots, G_k\}$ is the clause representation of $\neg G$. We will see later that the set-of-support restriction of resolution is complete.

The *input-restriction* of resolution (or just *input resolution*) requires that in each resolution step, one of the parent clauses has to be an "input", i.e. an element of the original clause set F. It is easy to see that an input resolution proof necessarily is a linear resolution proof. But in contrast to linear resolution, input resolution is not complete. The above discussed unsatisfiable clause set

$$F = \{\{A, B\}, \{A, \neg B\}, \{\neg A, B\}, \{\neg A, \neg B\}\}.$$

is a simple counter example. In this example, the first resolution step produces a clause with a single literal. Each further step produces then, by the input restriction, single element clauses. Therefore, the empty clause is not derivable by input resolution. But, we will later see that input resolution is complete when restricted to clause sets which contain only Horn clauses.

Another incomplete resolution restriction is *unit resolution*. Unit resolution is also complete for Horn clauses (see also Exercise 35). It is only allowed to produce a resolvent if at least one of the parent clauses is a unit, i.e. contains only a single literal. This resolution restriction has the advantage that the size of the produced resolvents decreases as compared with the parent clauses. Hence, unit resolution is working towards producing the empty clause which has size 0. The incompleteness of unit resolution can be seen by the same counter example as for input resolution, and this is not mere accident: It can be shown that a clause set has an input resolution refutation if and only if it has a unit resolution refutation (cf. Exercise 91).

We finally proceed to the *SLD-resolution* (SLD = linear resolution with selection function for definite clauses). This restriction is only defined for Horn clauses. This resolution restriction plays an important role in logic programming which will be discussed in more depth in the next chapter. SLD-resolutions are both input and linear resolutions which have a special form. The base clause must be a negative clause (a so-called *goal clause*), and in each resolution step, the side clause must be a non-negative input clause. (A non-negative Horn clause is also named a *definite clause* or a *program clause*).

For example, let $F = \{C_1, C_2, \ldots, C_n, N_1, \ldots, N_m\}$ be a set of Horn clauses where $C_1, C_2, \ldots, C_n$ are the definite clauses and $N_1, \ldots, N_m$ are the goal clauses. An SLD-resolution of the empty clause must then have the form, for a suitable $j \in \{1, \ldots, m\}$ and for a suitable sequence $i_1, i_2, \ldots, i_l \in \{1, \ldots, n\}$.

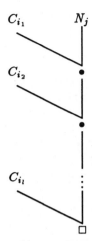

The clauses represented by dots, i.e. the "intermediate results", can only be negative clauses, because they result from resolution of a negative and a definite Horn clause. That means, SLD-resolutions are always N-resolutions. Furthermore, SLD-resolutions are set-of-support resolutions where the set-of-support is $\{N_1, \ldots, N_m\}$ (cf. Exercise 39). We will show that SLD-resolution is complete for Horn clauses.

Remark: In the abbreviation SLD (linear resolution with selection function for definite clauses), the additional aspect of a *selection function* is mentioned. In our present definition, we ignore this aspect of selection, but come back to this point in the investigations of Section 3.3. There, the presence of a selection function (which selects the next definite clause to be resolved with) is treated as combination of SLD-resolution with a special *strategy* (see the discussion at the beginning of this section). Here, we treat SLD-resolution as identical with *LUSH-resolution* (LUSH = linear resolution with unrestricted selection for Horn clauses).

All of the completeness proofs for these resolution restrictions are shown for the propositional case first, that is, for the ground instances of the predicate logic clauses. Just as in the proof of the general resolution theorem of the last section, the Lifting Lemma is used to convert resolution refutations for ground instances to resolution refutations for the original clauses in predicate logic. We have to check that the Lifting Lemma does not change the structure of a resolution proof. It is easily seen that a P-, N- etc. resolution is still a P-, N- etc. resolution after application of the Lifting Lemma. To prove completeness of a resolution restriction, we have to modify the proof of the resolution theorem in propositional logic (see Section 1.5) according to the respective restriction.

As preparation for the following proofs, we introduce the following notation. For a (propositional logic) clause set F and a literal L occuring in F, we let $F_{L=0}$ be the clause set which is obtained from F by canceling every occurence of L within the clauses of F, and for every occurence of $\overline{L}$ in a clause in F, the whole clause is eliminated from F. Similarly, $F_{L=1}$ is defined by interchanging the roles of L and $\overline{L}$. In other words, $F_{L=a}$, $a \in \{0, 1\}$, is obtained from F by fixing the assignment $\mathcal{A}(L) = a$ and performing obvious simplifications. From this, it is clear, that the unsatisfiability of F implies the unsatisfiability of $F_{L=0}$ and of $F_{L=1}$.

Theorem

The P-restriction of resolution is complete.

Proof: As observed above, it suffices to prove the theorem for propositional logic. Let F be an unsatisfiable set of clauses. By the compactness theorem, we can assume that F is finite. We show by induction on the number of different atomic formulas occuring in F, that the empty clause is deductable from F by P-resolution.

If $n = 0$, then $F = \{\Box\}$, and there is nothing to prove.

Now let $n > 0$ and assume that F contains n atomic formulas. Pick any one of those, say A. Then both clause sets $F_{A=0}$ and $F_{A=1}$ are unsatisfiable and contain at most $n - 1$ atomic formulas. By induction hypothesis, there are resolution refutations for $F_{A=0}$ and for $F_{A=1}$ satisfying the P-restriction. Now we insert the literal A in all those clauses in $F_{A=0}$ again, where it was canceled before, and also in all the respective resolvents. The above resolution of the empty clause from $F_{A=0}$ then turns into a resolution of $\{A\}$ from F. This is still a P-resolution since A is a positive literal. Next, we add resolution steps which resolve the so-obtained clause $\{A\}$ with every clause in F which contains $\neg A$. These resolution steps are also P-resolutions. Now we have all clauses from $F_{A=1}$ available. Therefore, we attach the P-resolution refutation building upon $F_{A=1}$, which exists by induction hypothesis, and obtain altogether a P-resolution refutation of F.
                                                                                    ∎

Theorem

The N-restriction of resolution is complete.

Proof: Swap in the above proof all occurences of "positive" by "negative", of A by $\neg A$, of $F_{A=0}$ by $F_{A=1}$ (and vice versa). ∎

Theorem

Linear resolution is complete. (More precisely: For every unsatisfiable clause set F there is a clause $C \in F$ such that the empty clause is linearly resolvable from F, based on C).

Proof: Let F be unsatisfiable. By the compactness theorem, we can assume that F is finite. Let F' be a minimally unsatisfiable subset of F (i.e. F' is an unsatisfiable subset of F, and every proper subset of F' is satisfiable. F' can be constructed from F by successively canceling clauses from F until any further canceling of a clause causes satisfiability of the resulting clause set).

Now we show that *every* clause in F' can be used as base clause to allow a linear resolution refutation. The proof is by induction on the number n of atomic formulas occuring in F'. Let C be an arbitrary clause in F'. If $n = 0$, then $F' = \{\square\}$ and $C = \square$. There is nothing to show.

Now we come to the induction step. If F' contains $n > 0$ atomic formulas, then we consider two cases.

Case 1: $|C| = 1$.

In this case, $C = \{L\}$ for some literal L. Then the clause set $F'_{L=1}$ is unsatisfiable and contains at most $n - 1$ different atomic formulas. Let F'' be a minimally unsatisfiable subset of $F'_{L=1}$. Then we claim that F'' must contain a clause C' such that $C' \cup \{\overline{L}\} \in F'$. That means, C' was obtained from a clause in F by canceling $\overline{L}$. Such a clause C' must exist in F'' because otherwise F'' would be a subset of $F' - \{C\}$, and therefore would be satisfiable (because F'' was chosen minimally unsatisfiable). By the induction hypothesis, there is a linear resolution of the empty clause from F'', based on C'. From this linear resolution proof we construct the desired linear resolution of the empty clause from F', based on $C = \{L\}$, as follows. The first resolution step resolves the base clause $C = \{L\}$ with $C' \cup \{\overline{L}\}$. Therefore, the resolvent is C'. Then we attach the above resolution refutation, but take the original clauses from F instead, i.e. possibly with the literal $\overline{L}$ which was canceled in F'. The literal $\overline{L}$ also appears in the respective resolvents, and instead of deducing the empty clause, we obtain $\{\overline{L}\}$ at the end of the linear chain. A final resolution step resolving $\{\overline{L}\}$ with the base clause $C = \{L\}$ gives the empty clause.

Case 2: $|C| > 1$.

In this case, we choose an arbitrary literal $L \in C$ and let $C' = C - \{L\}$. Then, $F'_{L=0}$ is unsatisfiable and C' is a clause in $F'_{L=0}$. We claim that $F'_{L=0}$ is satisfiable. To see this, let $\mathcal{A}$ be a model for $F' - \{C\}$. Then, $\mathcal{A}(C) = 0$, because $\mathcal{A}(F') = 0$, by unsatisfiability of F'. Therefore, $\mathcal{A}(L) = 0$, because $L \in C$. From this, we obtain $\mathcal{A}(F'_{L=0} - \{C\}) = 1$.

Now let F'' be a minimally unsatisfiable subset of $F'_{L=0}$. As we have just seen, F'' must contain C' (because canceling C' from F'' would cause satisfiability). We can apply the induction hypothesis on F''. Therefore there exists a linear resolution of the empty clause from F'', based on C'. In this resolution proof, we add the literal L at every place where it was canceled before (also in the respective resolvents). Then we obtain a linear resolution of $\{L\}$ from F', based on C.

Now we observe that $(F' - \{C\}) \cup \{\{L\}\}$ is unsatisfiable and $F' - \{C\}$ is satisfiable. Using Case 1, there exists a linear resolution of the empty clause from $(F' - \{C\}) \cup \{\{L\}\}$, based on $\{L\}$. Attaching this resolution proof behind the above contructed resolution which yields $\{L\}$, we obtain the desired resolution of the empty clause from F', based on C.   ∎

Exercise 87: Let F be the unsatisfiable clause set built up from the atomic formulas $A_1, \ldots, A_n$ such that F contains all $m = 2^n$ clauses of the form $\{B_1, B_2, \ldots, B_n\}$ with $B_i \in \{A_i, \neg A_i\}$. The usual resolution refutation (in form of a complete binary tree) has $m - 1$ resolution steps. Find recursions for the number of resolution steps constructed by the induction proofs for completeness of linear and of P-resolution. Compare this number with $m - 1$.

Theorem

The set-of-support restriction of resolution is complete.

Proof: This follows from the completeness of linear resolution. Let F be an unsatisfiable clause set, and let $T \subseteq F$ be a set-of-support, i.e., $F - T$ is satisfiable. A minimal unsatisfiable subset of F has to contain at least one clause $C \in T$, because $F - T$ is satisfiable. Using the previous proof, it follows that there is a linear resolution of the empty clause from F, based on C. This is also a set-of-support resolution with set-of-support $\{C\}$, and therefore also with set-of-support T.   ■

Exercise 88: Show that by combining two complete resolution restrictions, in general, one loses completeness. Give an example of an unsatisfiable clause set and two complete resolution restrictions (e.g. P-resolution and N-resolution) such that the empty clause cannot be derived by satisfying *both* restrictions.

Now we turn to the resolution restrictions which are incomplete in the general case. Here we obtain immediately the following theorem (cf. Exercise 35).

Theorem

Unit resolution is complete for the class of Horn clauses.

Proof: Since P-resolution is complete in the general case, it is also complete for the special case of Horn clauses. But positive Horn clauses must be units (i.e. consist of a single literal). Therefore, it immediately follows that unit resolution is complete for Horn clauses.   ■

Theorem

SLD-resolution is complete for the class of Horn clauses.

Proof: Let F be an unsatisfiable set of Horn clauses. Such a set must contain a negative clause (otherwise let $\mathcal{A}(A) = 1$ for every atomic formula A in F, then $\mathcal{A}$ would be a model for F). Furthermore, if F' is a minimally unsatisfiable subset of F, then there must be a negative clause C in F'. By completeness of linear resolution, there is a resolution of the empty clause from F' (hence from F), based on C. This linear resolution chain must have the form of an SLD-resolution. First, it is based on a goal clause, namely C. Further, all resolvents must be negative clauses, therefore the side clauses can only come from F, and must be definite clauses.   ■

Exercise 89: Prove the completeness of SLD-resolution for the class of Horn clauses *directly*, i.e. without refering to the completeness of linear resolution.

Hint: Imagine the process of the marking algorithm for Horn clauses discussed in Section 1.3. and "simulate" it backwards in terms of SLD-resolution steps. Another possibility is to use (a generalization of) Exercise 38.

Theorem

Input resolution is complete for the class of Horn clauses.

Proof: SLD-resolutions are also input resolutions.   ∎

Exercise 90: Show that the completeness of input resolution follows just as easily from the completeness of N-resolution.   ∎

Exercise 91: Show that for every clause set F, F has an input resolution refutation *if and only if* F has a unit resolution refutation.

Exercise 92: Show that resolution remains complete if no resolution step is allowed where one of the parent clauses is a tautology. A clause is a tautology if and only if it contains an atomic formula together with the complement of this atomic formula.

Exercise 93: If, in a resolution step, only *one* literal in each parent clause is used for unification, then we call this *binary* resolution. (In other words, in the definition of a resolvent in predicate logic, $m = n = 1$).

Show by a counter example that binary resolution in general is incomplete. Show further that binary resolution is complete for Horn clauses. Furthermore, for Horn clauses it remains complete if combined with any of the other complete resolution restrictions for Horn clauses.

(Actually, it is this combination of binary resolution ond SLD-resolution which plays a special role in the next chapter).

Remark: We have made an effort to use the possibilities which the semi-decidability of predicate logic offers; that is, the possibility of obtaining automated theorem proving procedures. But practical experience with such procedures (which might involve trickier techniques than those we discuss here) shows that they are not yet able to prove complicated theorems – or to substitute for mathematicians.

It might be the case that the approach is too general. If

$$M = \{F_1, F_2, \ldots, F_n\}$$

is an arbitrary set of axioms, and if it has to be tested whether the formula G is a theorem of the theory $Cons(M)$, in principle, resolution can do the job. One has to test whether $\{F_1, F_2, \ldots, F_n, \neg G\}$ is unsatisfiable. Very often, the interest is concentrated on very special theories, so that M is fixed, and only G varies.

In this case, it might be better to develop calculi which are directly tailored for the particular theory (but only applicable for that theory). Such a calculus, in a sense, incorporates more *knowledge* about the particular theory than the pure resolution calculus.

Chapter 3

LOGIC PROGRAMMING

3.1 Answer Generation

In this section we show that the execution of a program can be understood as the automated deduction of the empty clause from a given unsatisfiable clause set (possibly using the resolution refinements from Section 2.6). A further concept is needed: how to *generate an answer*, a result of the computation, from the resolution proof. A resolution proof as such shows only that the empty clause is derivable; an answer, in a sense, explains *how* it is obtained. The following ideas of extracting an answer from a resolution proof have their roots in the works of Green and Kowalski.

Suppose a satisfiable set of clauses F is given. We can interpret F as a (logic) program: In F certain predicates and function symbols occur; and by the clauses in F, certain assertions about the relationships of the predicate and function symbols are made. In a sense, the general context of the problem to be solved is specified.

Let us consider the following simple example (here we use for better understanding more intuitive names for the occuring symbols than P, f, and a, etc.).

$$F = \{ \{likes(Eve, Apples)\},$$
$$\{likes(Eve, Wine)\},$$
$$\{likes(Adam, x), \neg likes(x, Wine)\} \}$$

109

Here, *likes* is a binary predicate symbol, and *Eve, Apples, Wine, Adam* are constants. This clause set F can be interpreted as

> "Eve likes apples"
> "Eve likes wine"
> "Adam likes everybody who likes wine"

This can be thought of as a specification of the general problem context (in this case, not a very profound one). A *call* of this logic program might be given by the formula

$$G = \exists y \ likes(Adam, y).$$

This can be interpreted as the question: "Is there anybody whom Adam likes?", and furthermore – this is the aspect of answer generation – "Who is this?"

It is typical for such a formula as G which serves as the program call (or query) to contain an *existential* quantifier. We expect to receive from the system (the logic program evaluator) not only the answer *yes* or *no*, but in case of *yes* we additionally expect the presentation of such an object (or all such objects) whose existence is claimed.

In the terminology of the excursion in Section 2.3, the logic program F can be thought of as being an axiom system for some theory $Cons(F)$ (here the "theory of Adam and Eve"), and the question G to be answered corresponds to the question whether G is a theorem of $Cons(F)$. In other words, we want to know whether G is a consequence of F. We can verify this using resolution by checking whether there is a resolution refutation of $F \wedge \neg G$. We have

$$
F \wedge \neg G \equiv \{ \{ likes(Eve, Apples) \}, \\
\{ likes(Eve, Wine) \}, \\
\{ likes(Adam, x), \neg likes(x, Wine) \}, \\
\{ \neg likes(Adam, y) \} \}.
$$

A derivation of the empty clause from this clause set is given by the following diagram.

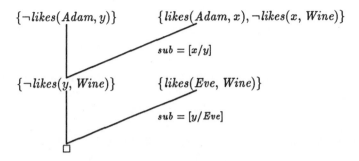

Indeed, the empty clause is derivable. Therefore, the clause set is unsatisfiable. This means that G is a consequence of F. Therefore, there is somebody whom Adam likes. But who? This can be seen from the substitutions which are performed on the variable in the query clause, namely y. The variable y is substituted by Eve in the second resolution step. Therefore the answer is Eve.

A possibility to make this substitution process transparent is to introduce an *answer predicate*. Instead of the clause $\{\neg likes(Adam, y)\}$ that stems from the query formula G, we now use

$$\{\neg likes(Adam, y), ANSWER(y)\}.$$

Now our aim is not to derive the empty clause, but to derive a clause that consists of (one or more) answer predicates only. In our example, we get:

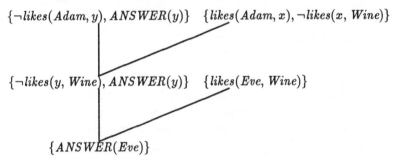

Another example with a more complex answer is the following.

$$\{ \{likes(Eve, Apples)\},$$
$$\{likes(Eve, Wine), likes(Lucy, Wine)\},$$
$$\{likes(Adam, x), \neg likes(x, Wine)\},$$
$$\{\neg likes(Adam, y), ANSWER(y)\} \}$$

An interpretation of this clause set is

"Eve likes apples"
"Eve or Lucy (or both) like wine"
"Adam likes everyone who likes wine"

and the query clause containing the answer predicate can be interpreted as

"Who does Adam like?"

An answer generation using resolution looks like this:

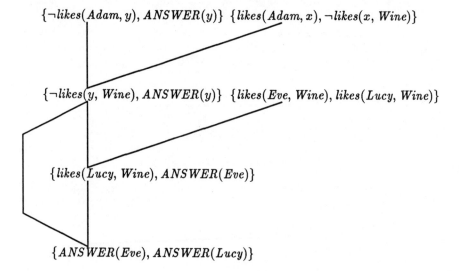

This means that Adam likes Eve *or* Lucy (or both). Notice that the resolution proofs can always be so arranged that they are *linear* resolutions, as above, with the query clause as base clause (cf. Section 2.6).

This more complicated situation that the answer consists of an *or* of two possibilities was enforced by the situation that the logic program contained the clause $\{likes(Eve, Wine), likes(Lucy, Wine)\}$. In general, this situation is possible whenever the logic program contains a clause with more than one positive literal, i.e., a clause that does not have the Horn form.

Another possibility for this to happen is when the query formula, after transforming it to clausal form, splits in more than one clause. Each of these clauses then contains the answer predicate. This is one of the reasons that PROLOG insists of using only Horn clauses, and only one query clause.

We will discuss in the next sections how the answer generation process for the special case of Horn clauses can be designed, using the resolution restrictions from Section 2.6 and further special evaluation strategies tailored for these restrictions.

Consider the query clause

$$\{\neg likes(Eve, z), ANSWER(z)\}$$

which is in words,

"Is there anybody or anything that Eve likes, and if so, output such an object z".

Here we obtain two possible deductions of the empty clause (or the pure answer clause) and therefore two possible answers.

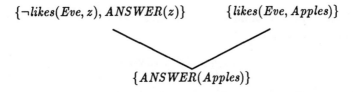

$$\{\neg likes(Eve, z), ANSWER(z)\} \qquad \{likes(Eve, Apples)\}$$

$$\{ANSWER(Apples)\}$$

and

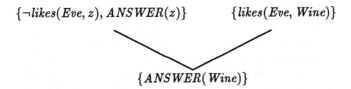

$$\{\neg likes(Eve, z), ANSWER(z)\} \qquad \{likes(Eve, Wine)\}$$

$$\{ANSWER(Wine)\}$$

This means that $z = Apples$ and $z = Wine$ are both possible answers. In this case, the *and*-combination of two answers is expressed by the situation that there are two possible resolution proofs.

Exercise 94: Modify the above example so that *exactly one* of Eve and Lucy likes wine. Then apply again the answer generation process.

Remark: Consider a set of clauses F (which can now also be called logic program) and let P be a predicate symbol occuring in F (e.g. *likes* in the above example). For simplicity, suppose P is binary. Then, a typical query formula has the form

$$G = \exists z\, P(t, z)$$

for some variable-free term t (for example, $t = Adam$). This formula G leads to the query clause $\{\neg P(t, z)\}$ or $\{\neg P(t, z), ANSWER(z)\}$. This situation can be thought of as if t is the actual *input parameter* and z is the *output parameter*. After evaluation of the logic program F with the query clause we expect the *result* of the computation to be substituted for the variable z. Our formalism permits an input parameter transfer and an output parameter transfer. Both are accomplished by the substitutions done by the unification algorithm (as a "subroutine" of the resolution algorithm). More precisely, the input parameter transfer is achieved by the substitutions for the variable(s) in the logic program F, and the output parameter transfer by the substitutions for the variables occuring in the query clause (i.e. in the answer predicate).

Notice that the same logic program F can be used with different query clauses, for example

$$\{\neg P(z', t'), ANSWER(z')\}.$$

Here the roles of input and output parameter are interchanged. Now the question is not what Eve likes (apples and wine), but who likes Eve (Adam). We say, the parameter passing process is *invertible*.

The next example is the *monkey-and-banana* problem. Here the function symbols are used as operators on a certain state space which characterizes the relative situation of the monkey, the chair, and the bananas. The aim of the computation is to find a series of applications of the available "operators" to transform a starting state into a desired end state (where the monkey has reached the bananas). Consider the following clauses

(1) $\{P(a, b, c, d)\}$

Interpretation: "In the start situation d the monkey is in position a, the banana is hanging above position b, and the chair is at position c." (Here a, b, c, d are constants).

(2) $\{\neg P(x, y, z, s), P(w, y, z, walk(x, w, s))\}$

"If, in some situation s, the monkey is in position x, then an application of the function $walk(x, w, s)$ has the effect that the monkey is afterwards in position w. In other words, the monkey is able to walk to any position." (Here, x, y, z, s, w are variables).

(3) $\{\neg P(x, y, x, s), P(w, y, w, push(x, w, s))\}$

"If the monkey is at the same position as the chair, namely x, then he can push it to any position w."

(4) $\{\neg P(x, y, x, s), P(x, y, x, climb(s))\}$

"If the monkey is at the chair, then he can climb the chair."

(5) $\{\neg P(x, x, x, climb(s)), Reach(grasp(climb(s)))\}$

"If the monkey has climbed the chair, and if the position of monkey, chair, and banana coincide, then the monkey can reach the banana by grasping it."

The clauses describe the problem context and are considered as the logic program. Now consider the question

$$\exists z \ Reach(z),$$

which means in words: "Is there a situation in which the monkey has reached the banana – and how to achieve it?" Again, we negate the question and transform it into clause representation, including the answer predicate, and obtain

(6) $\{\neg Reach(z), ANSWER(z)\}.$

A resolution proof of the pure answer answer clause is given by the following sequence $C_1, \ldots, C_5$ with

$C_1 \;=\; \{\neg P(x, x, x, climb(s)), ANSWER(grasp(climb(s)))\}$
      (resolvent of (5) and (6))
$C_2 \;=\; \{\neg P(x, x, x, s), ANSWER(grasp(climb(s)))\}$
      (resolvent of (4) and C_1)
$C_3 \;=\; \{\neg P(x, y, x, s), ANSWER(grasp(climb(push(x, y, s))))\}$
      (resolvent of (3) and C_2)
$C_4 \;=\; \{\neg P(x, y, z, s), ANSWER(grasp(climb(push(x, y, walk(x, z, s)))))\}$
      (resolvent of (2) and C_3)
$C_5 \;=\; \{ANSWER(grasp(climb(push(c, b, walk(a, c, d)))))\}$
      (resolvent of (1) and C_4)

Interpretation of the answer: "Starting from the situation d, walk from a to c, push the chair from c to b, climb the chair, and grasp the banana."

Exercise 95: Six coins are lying on the table in the following order

head   head   head   tail   tail   tail

In one move, two adjacent coins may be turned. We search for a sequence of moves which tranfers the coins into the situation

tail   head   tail   head   tail   head

Formulate a logic program to solve this puzzle.

Exercise 96: Three young women with their three jealous boy friends want to drive to the beach. They have a sports roadster available with two seats. How can they arrange the drives to the beach so that at no moment a woman is together with another man – except her own boy friend is present?

Exercise 97: Formulate the following puzzle in predicate logic clauses, and use the answer generation method to solve it:

Tom, Mike, and John are members of the alpine club. Each member of the alpine club is either skier or climber or both. No climber likes the rain and all skiers like the snow. Mike likes everything that Tom dislikes, and vice versa. Mike and John like the snow.

Is there a member of the alpine club who is climber but no skier, and who is this?

Exercise 98: Consider again the theorem proving example on group theory from Section 2.5. Use the answer generation method to find out subsequently how the right inverses have to be chosen.

3.2 Horn Clause Programs

In the following, we consider logic programs (sets of clauses) that are restricted to the Horn form. There exists a well established theory for logic programs of this form. The programming language PROLOG is based on Horn clauses. There are several reasons for this restriction to Horn form.

First, most of the mathematical theories seem to be axiomatizable in terms of Horn formulas (provided they are axiomatizable at all). Many examples in this book (e.g. the monkey-and-banana problem) turn out to be in Horn form. Therefore, imposing the Horn form restriction does not seem to be a real restriction in practice.

Second, allowing clauses that are not Horn leads to more complicated answer situations. We have discussed this in terms of some examples in the last section. This is one of the reasons why there is no developed theory of answer generation (or, logic programming) for the general case. (Notice, we did not prove any theorems about correctness or completeness of the answer generation process in the last section).

The third reason is efficiency. In the propositional logic case, we have seen there are efficient algorithms for testing satisfiability of Horn formulas (Section 1.3 and Exercise 35), in contrast to the exponential algorithms in the general case. Certain aspects of efficiency are still present when we consider the case of Horn formulas in predicate logic. (Although, the undecidability result from Section 2.3 is still valid, even for the special case of Horn formulas).

In particular, it is the completeness of SLD-resolution for Horn clauses that is attractive here since SLD-resolution proofs have the nature of a sequential computation. In a sense, the input of such an SLD-computation is the base clause (the query clause), and the computation is successful and leads to a result if the empty clause is derivable. In view of this computational (or procedural) interpretation of SLD-resolution proofs, we distinguish between the following types of Horn clauses.

Clauses that consist of a single positive literal are called *facts* in the following. Such a clause can be interpreted as the claim of a simple positive statement.

Procedure clauses have the form $\{P, \neg Q_1, \ldots, \neg Q_k\}$ where $P, Q_1, \ldots, Q_k$ are certain atomic formulas of predicate logic. The notation in PROLOG, namely

$$P :- Q_1, Q_2, \ldots, Q_k.$$

shows the character of an implication (cf. Exercise 3). The symbol $:-$

stands for an implication sign pointing to the left. Here, P is called the *procedure head*, and the sequence $Q_1, \ldots, Q_k$ is called the *procedure body*. A single Q_i is considered as a *procedure call*.

The intended meaning here is that to satisfy the procedure body P, it suffices to perform the procedure calls $Q_1, \ldots, Q_k$ successfully. We will see later that this conception is closely related with an SLD-resolution refutation.

Notice that facts can be considered to be special cases of procedure clauses (with $k = 0$, i.e. there is no procedure body).

A *Horn clause program* (or in the following simply *logic program*) consists of a finite set of facts and procedure calls. An element of a logic program is also called *program clause* or *definite clause*.

Finally, a logic program is *called* or *activated* by a *goal clause*. A goal clause (also called *query clause*) is a Horn clause too, but one containing negative literals only. Such a clause has the form $\{\neg Q_1, \neg Q_2, \ldots, \neg Q_k\}$, or in the PROLOG notation,

$$?-\ Q_1, Q_2, \ldots, Q_k.$$

Refering again to the intuitive interpretation mentioned above, this notation suggests that a goal clause is a sequence of procedure calls which is to be satisfied successfully.

In this context the empty clause $\square$ is called the *halting clause*. It can be considered to be the special case of a goal clause (with $k = 0$) where all procedure calls are successfully performed.

In each resolution step, it is required that the variables of the two parent clauses are being renamed so that they are disjoint (these are the substitutions s_1 and s_2 in the definition of resolution, cf. Section 2.5). Obviously, it suffices to rename the variables in only one of the parent clauses (i.e. set $s_1 = [\ ]$). In the following SLD-resolution refutations we assume that renamings are performed for the program clauses only (which are the side clauses in the terminology of SLD-resolution, cf. Section 2.6), not in the goal clauses. We call this a *standardized* SLD-resolution.

Example: Consider the following recursive definition of the addition (letting y' denote the successor of y):

$$x + 0 \ = \ x$$

$$x + y' \;=\; (x + y)'$$

Formulated in predicate logical clauses, we obtain:

(1) $\{A(x, 0, x)\}$

(2) $\{A(x, s(y), s(z)), \neg A(x, y, z)\}$

Here $A(x, y, z)$ means that $x + y = z$, and s represents the successor function. The clauses (1) and (2) constitute the logic program. A possible goal clause could be

$$\{\neg A(s(s(s(0))), s(s(0)), u)\},$$

in words: "Compute $3 + 2$, and deliver the result in the variable u". A standardized SLD-resolution proof is given by the following diagram (here, z' is a new variable, obtained by renaming).

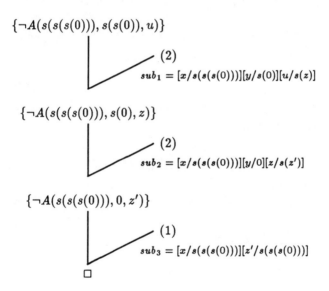

$\{\neg A(s(s(s(0))), s(s(0)), u)\}$

(2)

$sub_1 = [x/s(s(s(0)))][y/s(0)][u/s(z)]$

$\{\neg A(s(s(s(0))), s(0), z)\}$

(2)

$sub_2 = [x/s(s(s(0)))][y/0][z/s(z')]$

$\{\neg A(s(s(s(0))), 0, z')\}$

(1)

$sub_3 = [x/s(s(s(0)))][z'/s(s(s(0)))]$

$\square$

An answer, a result of the computation, can be obtained by applying the computed most general unifiers sub_1, sub_2, sub_3 to the original goal clause. We obtain

$$\{\neg A(s(s(s(0))), s(s(0)), u\}sub_1 sub_2 sub_3 =$$
$$\{\neg A(s(s(s(0))), s(s(0)), s(s(s(s(s(0)))))))\}.$$

To make it clearer what the result of the computation is, we can apply the substitution $sub_1 sub_2 sub_3$ directly to the variable u occuring in the goal clause.

$$
\begin{aligned}
u\, sub_1 sub_2 sub_3 &= s(z) sub_2 sub_3 \\
&= s(s(z')) sub_3 \\
&= s(s(s(s(s(0))))).
\end{aligned}
$$

In other words, the result is 5. This method of obtaining a result of the computation by applying the most general unifiers to the variable(s) in the goal clause is essentially identical with the method of introducing an answer predicate as discussed in the last section. In this example we can see that logic programs (in the pure form as described here) can only perform symbol manipulations, not arithmetical computations, as in standard programming languages. (We have obtained "s(s(s(s(s(0)))))" as result, not "5"). A concrete logic programming language (like PROLOG) should additionally support the possibility of *evaluating* arithmetical expressions (see the discussion about the *is* predicate in Section 3.4). At the moment we will not consider such non-logical aspects of a logic programming language.

Exercise 99: The logic program for addition descibed above can also be used for subtraction. How?

Exercise 100: Add to the addition program further clauses which allow one to compute the Fibonacci function. This is the function *fib* with

$$
\begin{aligned}
fib(0) &= 1 \\
fib(1) &= 1 \\
fib(n) &= fib(n-1) + fib(n-2) \text{ for } n \geq 2.
\end{aligned}
$$

Exercise 101: Formulate a different logic program for addition which is based on the following recursive presentation:

$$
\begin{aligned}
x + 0 &= x \\
x + y' &= x' + y.
\end{aligned}
$$

Compute again what the result of $3 + 2$ is.

Exercise 102: Ackermann's function is defined by the following equations.

$$\begin{array}{rcll} a(0, y) & = & y + 1 & \\ a(x, 0) & = & a(x - 1, 1) & \text{for } x > 0 \\ a(x, y) & = & a(x - 1, a(x, y - 1)) & \text{for } x, y > 0. \end{array}$$

For example, we have

$$\begin{array}{rcl} a(1, 2) & = & a(0, a(1, 1)) = a(0, a(0, a(1, 0))) \\ & = & a(0, a(0, a(0, 1))) = a(0, a(0, 2)) = a(0, 3) \\ & = & 4, \end{array}$$

whereas $a(4, 2)$ has more than 19000 decimal digits! Prove that this equational presentation of the function a is well defined, that is, each evaluation of $a(m, n)$ for $m, n \in \mathbb{N}$ ends in finitely many steps. Formulate a logic program to compute Ackermann's function!

The concepts introduced so far in terms of several examples will now be made more formal. Our aim is to define a rigorous formal semantics of such logic programs. In the following definition the "procedural" aspect of a logic program computation is emphasized.

Definition (procedural interpretation of Horn clause programs)

The procedural interpretation of Horn clause programs is given by the presentation of an abstract *interpreter* for such programs. A *configuration* of this interpreter is any pair (G, sub) where G is a goal clause and sub is a substitution.

Let F be a logic program (set of definite Horn clauses). The *transition relation* for configurations is then defined as follows.

$$(G_1, sub_1) \vdash_{\overline{F}} (G_2, sub_2)$$

if and only if G_1 has the form

$$G_1 = \{\neg A_1, \neg A_2, \ldots, \neg A_k\} \qquad (k \geq 1)$$

and there is a program clause

$$C = \{B, \neg C_1, \neg C_2, \ldots, \neg C_n\} \qquad (n \geq 0)$$

in F (after its variables have been renamed so that G_1 and C do not have a variable in common) such that B and A_i for some $i \in \{1, \ldots, k\}$ are unifiable. Let a most general unifier be the substitution s. Then G_2 has the form

$$G_2 = \{\neg A_1, \ldots, \neg A_{i-1}, \neg C_1, \ldots, \neg C_n, \neg A_{i+1}, \ldots, \neg A_k\}s$$

and sub_2 has the form

$$sub_2 = sub_1 s.$$

A *computation* of F on input $G = \{\neg A_1, \ldots, \neg A_k\}$ is a (finite or infinite) sequence of the form

$$(G, []) \vdash_{\overline{F}} (G_1, sub_1) \vdash_{\overline{F}} (G_2, sub_2) \vdash_{\overline{F}} \cdots$$

If the sequence is finite, and the last configuration of it has the form $(\Box, sub)$, then this computation is called *successful*, and in this case the formula $(A_1 \wedge \cdots \wedge A_k)sub$ is called the *result* of the computation.

It can be seen that a successful computation (restricted to the first component of the configurations) is simply a SLD-resolution refutation of $F \cup \{G\}$ where G is the base clause. Additionally, in the second component of a configuration, we keep track of the sequence of most general unifiers that have been used so far – similar to the answer predicate method in Section 3.1.

Notice that computations of Horn clause programs are *nondeterministic*, that is, each configuration can have more than one successor configuration. The possible computations from a given input G can be represented as a tree.

Not considering configurations that are different because of renaming of variables, this tree has bounded degree, but it might contain infinite paths.

Example: Consider the following logic program

$$F = \{\{P(x,z), \neg Q(x,y), \neg P(y,z)\},$$
$$\{P(u,u)\},$$
$$\{Q(a,b)\}\}$$

which in PROLOG notation is

$$P(x,z) :- Q(x,y), P(y,z).$$
$$P(u,u).$$
$$Q(a,b).$$

The goal clause $G = \{\neg P(v,b)\}$ (resp. ?– $P(v,b)$) as input leads to a non-successful computation

$(\{\neg P(v,b)\}, [\,])$

$\vdash_F$ $(\{\neg Q(v,y), \neg P(y,b)\}, [x/v][z/b])$

$\vdash_F$ $(\{\neg P(b,b),\}, [x/v][z/b][v/a][y/b])$

$\vdash_F$ $(\{\neg Q(b,y), \neg P(y,b)\}, [x/v][z/b][v/a][y/b][x/b][z/b])$

$\vdash_F$ $(\{\neg Q(b,b)\}, [x/v][z/b][v/a][y/b][x/b][z/b][y/b])$

which cannot be continued. Here the first, third, first, and second program clause have been used in the SLD-resolution steps.

There are also two successful computations with different results. These are

$(\{\neg P(v,b)\}, [\,])$

$\vdash_F$ $(\{\neg Q(v,y), \neg P(y,b)\}, [x/v][z/b])$ (with the 1. program clause)

$\vdash_F$ $(\{\neg P(b,b)\}, [x/v][z/b][v/a][y/b])$ (with the 3. program clause)

$\vdash_F$ $(\square, [x/v][z/b][v/a][y/b][u/b])$ (with the 2. program clause)

and

$(\{\neg P(v,b)\}, [\,])$

$\vdash_F$ $(\square, [v/b])$. (with the 2. program clause)

The first computation leads to the result

$$P(v, b)[x/v][z/b][v/a][y/b][u/b] = P(a, b),$$

and the second,

$$P(v, b)[v/b] = P(b, b).$$

Exercise 103: Describe all computations that are possible under the logic program

$$P(a, b).$$
$$P(x, y) :- P(y, x).$$

with the given goal clause $?- P(b, z).$

Exercise 104: Which type of program clause can be considered as a *recursive* procedure (in the sense of standard programming languages)?

Example: Consider the following logic program that might be part of a larger program for symbolic differentiation. This time we use the PROLOG notation throughout. Here in this example, x and 1 denote constants, A,DA,B,DB,C denote variables, and *diff* is a binary predicate symbol, wereas *sin*, *cos*, $+$, $*$ are function symbols. For better readability, we use infix notation for $+$ and $*$ (that is, we write $x + y$ instead of $+(x, y)$).

$$diff(x, 1).$$
$$diff(A + B, DA + DB) :- diff(A, DA), diff(B, DB).$$
$$diff(A * B, A * DB + B * DA) :- diff(A, DA), diff(B, DB).$$
$$diff(sin(A), cos(A) * DA) :- diff(A, DA).$$

This logic program formalizes the fact that the derivative of x is 1. The second, third, and fourth clause formalize the sum and the product rule, and the chain rule (with respect to the sin function).

Consider the goal clause

$$?- diff(x * sin(x), C).$$

In words: determine the derivative of $x * sin(x)$. The following computation is successful and yields the desired result $x*cos(x)+sin(x)$ (in the redundant form $x * (cos(x) * 1) + sin(x) * 1$).

$$(?- \ diff(x * sin(x), C), [\,])$$
$$\vdash_F \quad (?- \ diff(x, DA), diff(sin(x), DB), \ sub_1)$$
$$\vdash_F \quad (?- \ diff(sin(x), DB), \ sub_1 sub_2)$$
$$\vdash_F \quad (?- \ diff(x, DA), \ sub_1 sub_2 sub_3)$$
$$\vdash_F \quad (\square, \ sub_1 sub_2 sub_3 sub_4)$$

Here we have

$$sub_1 \ = \ [A/x][B/sin(x)][C/x * DB + sin(x) * DA]$$
$$sub_2 \ = \ [DA/1]$$
$$sub_3 \ = \ [A/x][DB/cos(x) * DA]$$
$$sub_4 \ = \ [DA/1]$$

Therefore,

$$C \, sub_1 sub_2 sub_3 sub_4 = x * (cos(x) * 1) + sin(x) * 1.$$

Exercise 105: Write a logic program that simplifies formulas by eliminating useless multiplicative factors that are 1, and additive terms that are 0. (This program could be combined with the above program for differentiation.) The program we are looking for should be able to perform the following computation. Using as input the goal clause

$$?- \ simplify(1 * A + (B + (0 + x)) * 1, C).$$

leads to the result

$$simplify(1 * A + (B + (0 + x)) * 1, A + (B + x)).$$

Exercise 106: Prove the following variation of the Lifting Lemma – tailored for Horn clause computations: If

$$(G \, sub', [\,]) \vdash_F \cdots \vdash_F (\square, sub)$$

is a computation of the logic program F with the input $Gsub'$, then there is another computation of F (of same length) on input G that has the form

$$(G,[]) \vdash_{\overline{F}} \cdots \vdash_{\overline{F}} (\Box, sub''),$$

such that for some suitable substition s,

$$sub'\,sub = sub''s.$$

Using the results from Section 2.6, it is immediately clear that $F \cap \{G\}$ is unsatisfiable if and only if there exists a successful computation of F with input G. This follows from the completeness of SLD-resolution (also, Exercise 93), and involves the logical aspect of Horn clause computations. But regarding the result of the computation, up to this point we are not able to make a statement about the correctness of such computational results. Neither do we know what the possible range of computation results is that might occur at the endpoints of the nondeterministic computation paths.

Such a statement is justified by the following theorem which can be understood as a strengthening of the correctness and completeness results for SLD-resolution obtained in sections 2.5 and 2.6. The theorem says that the obtainable computation results are as general as possible, that means they contain as many variables as possible.

Theorem (Clark)

Let F be a Horn clause program and let $G = ?\!- A_1, \ldots, A_k$ be a goal clause.

1. *(correctness property)* If there is a successful computation of F with input G, then every ground instance of the result $(A_1 \wedge \cdots \wedge A_k)sub$ is a consequence of F.

2. *(completeness property)* If every ground instance of $(A_1 \wedge \cdots \wedge A_k)sub'$ is a consequence of F, then there exists a successful computation of F with input G with the result $(A_1 \wedge \cdots \wedge A_k)sub$ such that for a suitable substitution s,

$$(A_1 \wedge \cdots \wedge A_k)sub' = (A_1 \wedge \cdots \wedge A_k)subs.$$

Proof: 1. The proof is by induction on the length n of the computation.

Induction Base: For $n = 0$ we have $G = \Box$ and $sub = [\,]$, and thus there is nothing to prove.

Induction Step: Let $n > 0$ and consider a typical computation of length n,

$$(G, [\,]) \vdash_{\mathrm{F}} (G_1, sub_1) \vdash_{\mathrm{F}} \cdots \vdash_{\mathrm{F}} (\Box, sub_1 \ldots sub_n).$$

Here, $sub_1, \ldots, sub_n$ are the most general unifiers provided by the unification algorithm. Let the goal clauses G and G_1 have the form:

$$G = ?\text{--} \ A_1, \ldots, A_{i-1}, A_i, A_{i+1}, \ldots, A_k \qquad (k \geq 1)$$

and

$$G_1 = ?\text{--} \ (A_1, \ldots, A_{i-1}, C_1, \ldots, C_l, A_{i+1}, \ldots, A_k) sub_1.$$

There must be a program clause in F of the form that was used for this first SLD-resolution step,

$$B \ :\text{--} \ C_1, \ldots, C_l \qquad (l \geq 0)$$

such that $\{B, A_i\}$ is unifiable with some most general unifier sub_1. Now consider the following computation of length $n - 1$:

$$(G_1, [\,]) \vdash_{\mathrm{F}} \cdots \vdash_{\mathrm{F}} (\Box, sub_2 \ldots sub_n).$$

By induction hypothesis, every ground instance of

$$(A_1 \wedge \cdots \wedge A_{i-1} \wedge C_1 \wedge \cdots \wedge C_l \wedge A_{i+1} \wedge \cdots \wedge A_k) sub_1 \ldots sub_n$$

is a consequence of F. In particular, the subformula

$$(C_1 \wedge \cdots \wedge C_l) sub_1 \ldots sub_n$$

is a consequence of F. Since $B \ :\text{--} \ C_1, \ldots, C_l$ is a program clause in F, and by the observation that

$$B sub_1 \ldots sub_n = A_i sub_1 \ldots sub_n,$$

it follows that every ground instance of $A_i sub_1 \ldots sub_n$ is a consequence of F, and hence also every ground instance of

$$(A_1 \wedge \cdots \wedge A_i \wedge \cdots \wedge A_k) sub_1 \ldots sub_n$$

is a consequence of F.

2. Let $x_1, \ldots, x_m$ be all the variables occuring in $Gsub'$, and let $a_1, \ldots, a_m$ be new constants which did not appear yet. Define

$$G' = Gsub'[x_1/a_1] \ldots [x_m/a_m].$$

By hypothesis, $F \cup \{G'\}$ is unsatisfiable, and by the completeness of SLD-resolution (cf. Section 2.6, also Exercise 93), there is a successful computation of F of the form

$$(G', [\,]) \vdash_{\mathbf{F}} \cdots \vdash_{\mathbf{F}} (\square, sub_1 \ldots sub_n).$$

Since G' does not contain any variables, $G' = G'sub_1 \ldots sub_n$. (That is, the substitutions in $sub_1 \ldots sub_n$ concern variables in the program clauses only.) Now we substitute $x_1, \ldots, x_m$ for $a_1, \ldots, a_m$ in this computation, and we obtain

$$(Gsub', [\,]) \vdash_{\mathbf{F}} \cdots \vdash_{\mathbf{F}} (\square, sub'_1 \ldots sub'_n).$$

Here, except for the above substitution, $sub'_1, \ldots, sub'_n$ is identical with $sub_1, \ldots, sub_n$. Therefore we have

$$Gsub' = Gsub'sub'_1 \ldots sub'_n.$$

Using the Lifting Lemma (cf. Section 2.5 and Exercise 106), the above computation can be transformed into another computation of the same length,

$$(G, [\,]) \vdash_{\mathbf{F}} \cdots \vdash_{\mathbf{F}} (\square, sub''_1 \ldots sub''_n).$$

Here $sub''_1, \ldots, sub''_n$ are most general unifiers provided by the Lifting Lemma. Then, for some suitable substitution s,

$$sub'sub'_1 \ldots sub'_n = sub''_1 \ldots sub''_n s.$$

Therefore, letting $sub = sub''_1 \ldots sub''_n$, we obtain

$$(A_1 \wedge \cdots \wedge A_k)sub' = (A_1 \wedge \cdots \wedge A_k)subs,$$

which completes the proof.   ∎

Next, we want to clarify what the *semantics* of a logic program is. As in standard (operational) programming languages, there are different approaches. First we give the definition of an *interpretative* or *procedural* semantics. This approach focuses on the idea that a logic program is (or induces) a parallel and nondeterministic *process*. The semantics of a logic program (together with a given goal clause as input) is the set of potential computational results of this process (with the given input). To simplify (and standardize) matters, we restrict ourself in the following to the *ground instances* of the computational results.

Definition (procedural semantics)

Let F be a logic program and G a goal clause. The *procedural semantics* of (F, G) is defined by the set of ground instances of the computation results of F on input G which the abstract logic program interpreter can produce. This is symbolically,

$$S_{proc}(F, G) = \{ \, H \mid \text{there is a successful computation of } F \text{ on}$$
$$\text{input } G \text{ such that } H \text{ is a ground instance of the}$$
$$\text{computation result} \, \}$$

Exercise 107: Show in detail what the procedural semantics of

$$P(a, a).$$
$$P(a, b).$$
$$P(x, y) :- P(y, x).$$

with the given goal clause

$$?- P(a, z), P(z, a).$$

is.

A second, quite different approach to define a semantics of logic programs starts out from the idea that the "meaning" (the *denotation*) of a logic program F – together with a given goal clause $G = ?- A_1, \ldots, A_k$ is the set of ground instances of $(A_1 \wedge \cdots \wedge A_k)$ which are consequences of F. This *model theoretic* approach is similar to the assignment of a theory $Cons(F)$ to a formula F (see Section 2.4). The theory $Cons(F)$ associated with the axiom system F can be thought of being the model theoretic semantics of F.

In contrast to the above idea that the logic program induces a dynamic process, we have here the idea of a static *data base*. The semantics of a logic program is declared as the set of formulas which is explicitly and implicitly represented by the program, namely everything which follows from it.

Definition (model theoretic semantics)

The *model theoretic semantics* of a logic program F and a given goal clause $G = ?- A_1, \ldots, A_k$ is the set of ground instances of $(A_1 \wedge \cdots \wedge A_k)$ that are consequences of F. In symbols,

$$\mathcal{S}_{mod}(F, G) = \{H \mid H \text{ is a ground instance of } (A_1 \wedge \cdots \wedge A_k)$$
$$\text{and } H \text{ is a consequence of } F\}.$$

Exercise 108: Find out what the model theoretic semantics of the example in Exercise 107 is.

The following theorem asserts that procedural and model theoretic semantics are equivalent. This can be understood as a reformulation of Clark's Theorem.

Theorem

For all Horn clause programs F and goal clauses G,

$$\mathcal{S}_{proc}(F, G) = \mathcal{S}_{mod}(F, G).$$

Proof: ($\subseteq$) Let $H \in \mathcal{S}_{proc}(F, G)$. Then there is a successful computation of F of the form

$$(G, []) \vdash_{F} \cdots \vdash_{F} (\square, sub)$$

such that H is a ground instance of $(A_1 \wedge \cdots \wedge A_k)sub$. By Clark's Theorem (part 1), it follows that H is a consequence of F. Therefore, $H \in \mathcal{S}_{mod}(F, G)$.

($\supseteq$) Let $H \in \mathcal{S}_{mod}(F, G)$. Then H is a ground instance of $(A_1 \wedge \cdots \wedge A_k)$ and H is a consequence of F. By Clark's Theorem (part 2), it follows that there is a computation of F of the form

$$(G, [\,]) \vdash_{\overline{F}} \cdots \vdash_{\overline{F}} (\Box, sub)$$

such that H is an instance (in this case, a ground instance) of $(A_1 \wedge \cdots \wedge A_k)sub$. Therefore, $H \in \mathcal{S}_{proc}(F, G)$. ∎

Exercise 109: We can associate with each logic program F a function Op_F that maps sets of atomic formulas into sets of atomic formulas.

$Op_F(M) = \{A' \mid$ there exists a program clause C in F of the form $\{A, \neg B_1, \ldots, \neg B_k\}$, $k \geq 0$ such that $\{A', \neg B'_1, \ldots, \neg B'_k\}$ is a ground instance of C and $B'_1, \ldots, B'_k$ is in $M\}$.

Let $Op_F^0(M) = M$ and $Op_F^{n+1}(M) = Op_F(Op_F^n(M))$ for $n \geq 0$.

Prove that

$$Fp_F = \bigcup_{n \geq 0} Op_F^n(\emptyset)$$

is the *least fixpoint* of the operator Op_F (with respect to $\subseteq$).

The *fixpoint semantics* of F with given goal clause $G = ?- A_1, \ldots, A_k$ is defined as

$\mathcal{S}_{fixpoint}(F, G) = \{H \mid H$ is a ground instance of $(A_1 \wedge \cdots \wedge A_k)$ and for all atomic formulas A in H, $A \in Fp_F\}$.

Prove that $\mathcal{S}_{fixpoint}(F, G) = \mathcal{S}_{proc}(F, G)$.

3.3 Evaluation Strategies

Logic programs are *nondeterministic*, i.e. after each computation step there can be more than one possibility for continuing the computation. For every configuration (G, sub) there can exist finitely many configurations

$(G_1, sub_1), (G_2, sub_2), \ldots, (G_k, sub_k)$ such that for $i = 1, 2, \ldots, k$, (G, sub) $\vdash_{\overline{F}} (G_i, sub_i)$.

Whenever nondeterministic programs have to be run on a real computer operating deterministically and sequentially, this nondetermism has to be resolved in as efficient a way as possible. What is needed here is some *evaluation strategy* which determines in which order the nondeterministic computation steps have to be performed.

Looking more closely, it can be seen that the nondeterminism in logic programs occurs in two different forms: We distinguish in the following between type 1 nondeterminism and type 2 nondeterminism.

Suppose, we have already selected a particular literal (i.e. a procedure call) in the goal clause which is to be unified with some procedure head of some program clause. If there are several such program clauses which can be used to produce resolvents, we call this *type 1 nondeterminism*.

Example: Consider the goal clause $?- A, B, C$. Suppose B is selected as the next procedure call to be performed. Suppose the logic program contains the program clauses

$$B :- D.$$
$$B.$$
$$B :- E, F.$$

Then this situation results in three potential SLD-resolvents, that is, in three new goal clauses:

$$?- A, D, C.$$
$$?- A, C.$$
$$?- A, E, F, C.$$

From these three possible continuations of the computation, only one, if any, might be successful. Furthermore, even if there are several successful computations, they might lead to different results. This freedom in the choice of the next program clause constitutes the type 1 nondeterminism.

If the goal clause consists of n literals (i.e. procedure calls), then each of these n literals can be used for unification in the next resolution step. This gives $n!$ many ways of evaluating such a goal clause. This freedom in the choice of the literal in the goal clause constitutes the type 2 nondeterminism.

Let us consider the above example. We describe the situation by a tree which expresses both types of nondeterminism.

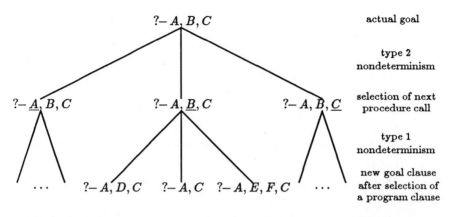

Next we show that the type 2 nondeterminism is not relevant and can be evaluated in *any* order. Every evaluation strategy concerning type 2 nondeterminism leads to the same computation results (so called "don't care" nondeterminism). That is, one loses no generality by fixing some special evaluation strategy. E.g., at the branching points for the type 2 nondeterminism one can follow the leftmost branch only and ignore the rest of the branches.

To justify this, we first show in a lemma that the evaluation order of procedure calls is not relevant, and can be swapped without changing the computation result.

Swapping Lemma

Consider two successive SLD-resolution steps

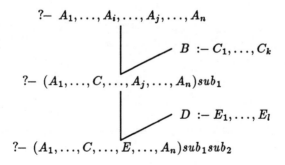

Here C stands for $C_1, \ldots, C_k$ and E stands for $E_1, \ldots, E_l$. Then the order of the resolution steps can be swapped:

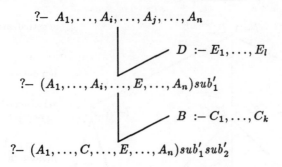

$$?-\ A_1, \ldots, A_i, \ldots, A_j, \ldots, A_n$$

$$D\ :-\ E_1, \ldots, E_l$$

$$?-\ (A_1, \ldots, A_i, \ldots, E, \ldots, A_n)sub'_1$$

$$B\ :-\ C_1, \ldots, C_k$$

$$?-\ (A_1, \ldots, C, \ldots, E, \ldots, A_n)sub'_1 sub'_2$$

Further, $sub_1 sub_2$ is identical with $sub'_1 sub'_2$ (except for possible renamings of variables).

Proof: First we have to show that the SLD-resolution steps can be performed in swapped order.

Observe that $A_j sub_1 sub_2 = D sub_2 = D sub_1 sub_2$, since sub_1 does not affect any variables in D. Therefore, A_j and D are unifiable, and the first resolution step can be performed. Let sub'_1 be a most general unifier of A_j and D. Since $sub_1 sub_2$ is a unifier of $\{A_j, D\}$, there is a substitution s such that $sub_1 sub_2 = sub'_1 s$.

Further, we have $Bs = B sub'_1 s = B sub_1 sub_2 = A_i sub_1 sub_2 = A_i sub'_1 s$. (The first equality is true because sub'_1 does not affect any variables in B). Therefore, $\{B, A_i sub'_1\}$ is unifiable using s. Hence the second resolution step can be performed. Let sub'_2 be a most general unifier.

It remains to show that $sub_1 sub_2$ and $sub'_1 sub'_2$ are (essentially) identical. We show that there are substitutions s' and s'' such that $sub_1 sub_2 = sub'_1 sub'_2 s'$ and $sub'_1 sub'_2 = sub_1 sub_2 s''$.

Since sub'_2 is a most general unifier of $\{B, A_i sub'_1\}$, and by the fact that $Bs = A_i sub'_1 s$, there is a substitution s' with $s = sub'_2 s'$. Therefore, $sub_1 sub_2 = sub'_1 s = sub'_1 sub'_2 s'$.

Next we observe that $A_i sub'_1 sub'_2 = B sub'_1 sub'_2$, and by the fact that sub_1 is a most general unifier of $\{A_i, B\}$, there is a substitution s_0 such that $sub'_1 sub'_2 = sub_1 s_0$. Now we have $A_j sub_1 s_0 = A_j sub'_1 sub'_2 = D sub'_1 sub'_2 = D sub_1 s_0 = D s_0$. (The last equality holds since sub_1 does not affect the variables in D). This means that s_0 is a unifier of $\{A_j sub_1, D\}$. By the fact that sub_2 is a most general unifier of $\{A_j sub_1, D\}$, there is a substitution s'' such that $s_0 = sub_2 s''$. Put together, we obtain $sub'_1 sub'_2 = sub_1 s_0 = sub_1 sub_2 s''$, what was to be shown. ∎

Definition

A computation of a logic program is called *canonical* if in each computation step the *first* literal (i.e. the literal at the leftmost position) in the goal clause is used for the resolution. (Notice that we consider clauses here as sequences, not as sets, and presume an ordering of the literals in the clause).

Theorem

Let $(G, [\,]) \vdash_F \cdots \vdash_F (\square, sub)$ be a successful computation of the logic program F. Then there exists a successful computation of F of same length which is canonical and which obtains the same computation result.

Proof: We assume that the given computation is canonical up to the ith computation step $(i \geq 0)$. Now we show how to transform this computation into one of same length with the same computation result that is canonical up to step $i + 1$.

Assume that after i computation steps the configuration (H, sub) is reached. Let $H = ?- A_1, \ldots, A_k$. The $(i + 1)$-th step is not canonical. Therefore, some literal A_l, $l > 1$, is used for resolution in this step, whereas this occurence of literal A_1 (or some instance of it) is used for resolution in some later computation step, say j $(j > i+1)$. Now we apply the Swapping Lemma to the pairs of computation steps $(j - 1, j)$, $(j - 2, j - 1)$, $\ldots$, $(i+1, i+2)$ and obtain a computation which is canonical up to length $i+1$. (It is not wrong to be reminded of Bubble-Sort at this point). Successive application of the above procedure makes the whole computation canonical.

∎

The theorem asserts that it is allowed to restrict ourselves to computations which are canonical. (In other words, this type of restriction is *complete*, cf. Section 2.6). Of course, we still have to deal with the type 1 nondeterminism. This theorem explains in retrospect what the S (for *selection function*) in the abbreviation SLD exactly means. Under every selection strategy regarding the type 2 nondeterminism (for example, the "left to right" strategy adopted in canonical computations) the SLD-resolution stays complete (for the class of Horn formulas).

Observe that canonical SLD-refutations (or better: SLD-computations) operate like *nondeterministic pushdown automata*: The content of the pushdown is the actual goal clause $?- A_1, A_2, \ldots, A_k$ where A_1 is the top element.

In each computation step, the top element A_1 is popped from the stack, and the procedure body $C_1, \ldots, C_n$ of some program clause $B :- C_1, \ldots, C_n$ is pushed on the stack, provided A_i and B are unifiable. In contrast to pushdown automata, the most general unifier *sub* provided by this unification is applied to the whole pushdown, so that the next goal clause has the form

$$?- C_1 sub, \ldots, C_n sub, A_2 sub, \ldots, A_k sub.$$

A further aspect is that we keep track of the evaluated most general unifiers in the second component of configurations so that we are able to specify the computation result.

We represent the canonical computations of a logic program F on input G as a tree where the root is labeled by the start configuration $(G, [\,])$. The sons of a father node labeled with (G', sub) are labeled with the successor configurations of (G', sub) according to a canonical computation. For better readability we often leave out the second components of configurations, and just label a node by the corresponding goal clause.

Example: Consider the logic program

$$
\begin{array}{lll}
1. & Q(x, z) & :- Q(y, z), R(x, y). \\
2. & Q(x, x). & \\
3. & R(b, c). &
\end{array}
$$

and the goal clause $?- Q(x, c)$. Then we obtain the following computation tree (where we have additionally labeled the edges by the number of the selected program clause):

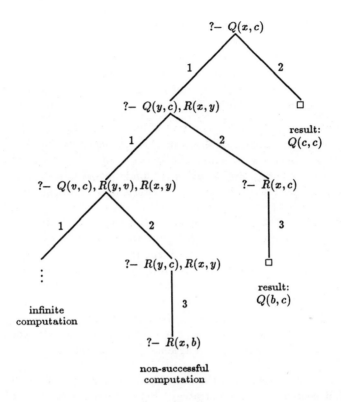

This tree has two successful computations with the different computation results $Q(b, c)$ and $Q(c, c)$. Further, there is a finite non-successful computation (i.e. non-extendable to the empty clause). If in every step, the first program clause is used for resolution, we obtain an infinite computation, where the goal clauses become longer by one literal in each step.

Exercise 111: In the approach taken above, clauses are sequences of literals, not sets of literals. So, identical elements in a sequence do not "melt" into a single element as for sets. It can happen now that pure propositional logic programs have infinite computations. Find an example!

Next we want to consider the possibilities of deterministically evaluating canonical computation trees. We have eliminated the type 2 nondeterminism by introducing canonical computations. Type 1 nondeterminism seems

to be more sensitive to changes or restrictions in the evaluation order. The
above examples show that we cannot just cut off certain parts of the tree.
The (type 1) nondeterministic choice of the next program clause is very
critical and determines whether a successful or non-successful (possibly infinite) computation is obtained, and in case of a successful computation,
the choices of program clauses determines the computation result.

Since there is (apparently) no way of eliminating the type 1 nondeterminism as we did in case of the type 2 nondeterminism, a deterministic
evaluation strategy has to search the whole computation tree (for a given
input) – at least until a first solution is found. In the following we will discuss two principle possibilities: *breadth-first search* and *depth-first search*.

In breadth-first search the search in the tree is performed so that all
nodes on depth t are visited (e.g. from left to right) before any node on
depth $t+1$ is visited ($t = 0, 1, 2, \ldots$). It should be clear that every successful
computation in the computation tree of a logic program can be found this
way after finitely many steps. In other words, the breadth-first search
evaluation strategy is *complete*. But this completeness is paid for in the
form of computation time and space: To reach the nodes in the computation
tree of depth t, the breadth-first search strategy needs to visit exponentially
many nodes (in t) – provided the tree consists not only of a single path.

Standard interpreters for the programming language PROLOG use the
depth-first search evaluation strategy. Here, starting from the root of the
tree, the subtrees are visited in some fixed order (from left to right) recursively. In contrast to breadth-first search the search goes into the depth of
the tree first. Whenever a node is reached that has no sons to search left
the search returns to the father node (backtracking) and continues with the
next brother node (if any).

For example, breadth-first search evaluates the tree

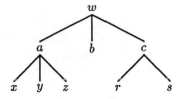

in the order

$$w, a, b, c, x, y, z, r, s$$

whereas depth-first search in the order

$$w, a, x, y, z, b, c, r, s.$$

The following algorithm realizes the depth-first backtracking evaluation strategy, as done in PROLOG.

PROLOG's Evaluation Strategy

Given: Logic program $F = (C_1, C_2, \ldots, C_n)$, where
$C_i = B_i :- D_{i,1}, \ldots, D_{i,n_i}$, and goal clause $G = ?- A_1, \ldots, A_k$.

The main routine consists of

    success := **false**;
    evaluate($G, [\,]$);
    if success **then** write('yes') **else** write('no');

and the recursive procedure *evaluate* works as follows:

    procedure evaluate(G : goalclause ; *sub* : substitution);
    var i : **integer**;
    begin
      if $G = \square$ **then**
        begin
        $H := (A_1 \wedge \cdots \wedge A_k)sub$;
        write('RESULT:',H);
        success := **true**
      end
      else {assume G has the form $G = ?- E_1, \ldots, E_k$}
        begin
        $i := 0$;
        while $(i < n)$ **and not** success **do**
          begin
          $i := i + 1$;
          if $\{E_1, B_i\}$ is unifiable using most general unifier s
            (where the variables in B_i have been renamed first)
            then
               evaluate(?- $(D_{i,1}, \ldots, D_{i,n_i}, E_2, \ldots, E_k)s$, *subs*)

        end
      end
  end;

Exercise 112: In real PROLOG systems, after finding a successful computation, the user is asked whether he wants to see more. Modify the above algorithm accordingly.

Observe that PROLOG's depth-first evaluation strategy *might* be faster than the breadth-first strategy. Consider a computation tree which has a successful computation of length t, and this computation is located at the very left of the tree. In this case depth-first will find it in about t search steps whereas breadth-first will still need exponentially many steps.

If the solution is at the very right of the tree, then depth-first is at least as inefficient as breadth-first. Even worse: computation trees might contain infinite paths (see the above example), therefore it is possible that the depth-first evaluation gets into an infinite loop before it ever reaches the successful computation. In other words, depth-first is – although sometimes more efficient than breadth-first – an incomplete evaluation strategy.

We summarize this discussion in the following theorem.

Theorem

The breadth-first evaluation strategy for logic programs is complete. The depth-first evaluation strategy is incomplete.

Exercise 113: One might try the following solution towards enforcing completeness of the depth-first search strategy. After the logic program F and the goal clause G are given, the program clauses in F are ordered in "some appropriate way". After this preprocessing step, the depth-first search strategy starts. (The hope is that this might turn all infinite computation paths in the tree to the right of the potential solution path).

Show that this approach fails, since there are examples for F and G such that $F \cup \{G\}$ is unsatisfiable, and therefore, by completeness of SLD-resolution, a successful computation exists, but the depth-first evaluation strategy will go into an infinite loop under *every* arrangement of the program clauses.

Hint: Consider the logic program for the monkey-and-banana problem.

By the advantages in efficiency, most PROLOG interpreters stick to this incomplete depth-first evaluation strategy. In a way, the problem is passed off to the programmer. He has to be aware of the PROLOG evaluation mechanism and has to plan the arrangement of his program clauses carefully. Even this will not help in some cases (see Exercise 113). To overcome this difficulty, PROLOG provides certain non-logical operators, like the *cut*, that influence PROLOG's evaluation order, see the next section). This is in conflict with the ideal conception of logic programming: the "programmer" should only provide the logical problem specification, whereas the system takes care of the algorithmic evaluation of the problem specification.

Kowalski introduced the equation

$$\text{algorithm} = \text{logic} + \text{control}$$

in the sense that algorithms always contain implicitly two components: a *logic component* which specifies the knowledge about the problem in question, and a *control component* which constitutes the solution strategy for the problem. In usual programming languages both components are strongly mixed and not separable, whereas in a logic programming language the program should only embody the logic component, and the control component should be a matter of the system, that carries out the evaluation algorithm.

This ideal case described above is certainly not yet realized by existing PROLOG implementations, using the depth-first evaluation strategy. On the other hand, the breadth-first strategy is hopelessly inefficient. One has to compromise on the ideal concept of logic programming (total separation of logic and control component) and on efficiency.

3.4 PROLOG

This section is not intended to be a PROLOG manual. We only wish to demonstrate some of the aspects that are relevant when stepping from the pure logic programming concept – as discussed in the last sections – to a real-life programming language, like PROLOG. PROLOG was developed in the seventies by a research group around A. Colmerauer in Marseille, France.

First, there must be syntactical conventions to enable a distinction between the different syntactical entities that occur in a logic program (clauses, variables, function and predicate symbols, logical operators). For example, in PROLOG one has to use upper-case symbols to identify variables, whereas function symbols and predicate symbols are written in lower-case. Furthermore, every clause must end with a period. In this section we will adopt these conventions.

In a practical programming language, concepts are needed that allow one to read data from some external device, like the keyboard or some file. The program must be able to write on the screen or into some file. In PROLOG, these tasks are accomplished by providing certain *system predicates* like *read* and *write* that cannot be modified by the user. From the logical standpoint, these predicates do not have a meaning (they immediately evaluate to **true** (or 1)), but they produce *side effects*, like writing a symbol on the screen or into a file.

If the PROLOG programmer uses such system predicates, it becomes necessary that he/she is aware of the evaluation strategy of PROLOG. A goal clause like

$$?-\ read(X),\ compute(X,Y),\ write(Y).$$

can be evaluated in a sensible way only from left to right. This is in contrast to the theoretical investigations of Section 3.3 where it was shown that such a logic program (without side effects) can also be evaluated from right to left (cf. the Swapping Lemma).

Other system predicates provided in PROLOG enforce certain instantiations of variables that deviate from the unification algorithm. An example is the predicate *is*. For example, if the PROLOG system find the clause $is(X, Y * 5)$ (or in infix notation: X *is* $Y * 5$) and Y is already bound (i.e. instantiated) to the constant 7, then X will be instantiated to 35. By this concept it is possible to perform arithmetical computations in PROLOG.

Example (cf. Exercise 100):

$$fib(0, 1).$$
$$fib(1, 1).$$
$$fib(X, Y) \quad :- \quad X_1 \text{ is } X - 1, X_2 \text{ is } X - 2,$$
$$fib(X_1, Y_1), fib(X_2, Y_2), Y \text{ is } Y_1 + Y_2.$$

Exercise 114: Using the *is* predicate, compute the factorial function by a PROLOG program.

Using the system predicate *is*, the invertibility of the parameter passing mechanism is lost. The above program for the Fibonacci function can only be used in a way that the first parameter is the input parameter and the second is the output parameter.

A further aspect of PROLOG implementations is that functions and predicates can either be written in prefix notation (e.g. $+(5,7)$) or infix notation ($5+7$). Further, PROLOG does not make a real distinction between function and predicate symbols. This goes even so far that the (logical) symbol $:-$ that stands for the implication sign is handled like a special system predicate/function, written in infix notation, that needs a special evaluation by the PROLOG system. The non-distinction between predicate and function symbols has the consequence that clauses and terms have to be considered as the same syntactical objects. Therefore, PROLOG allows variables on clause positions, and allows them to be instantiated with clauses. Therefore, a PROLOG program is able to manipulate its own "data base" (by the system predicates *assert* and *retract*).

More complex data structures are expressed in PROLOG by using nestings of terms. For example, the term

$$cons(a, cons(b, cons(c, nil)))$$

denotes a list consisting of the three elements a, b, and c. Here, the constant *nil* denotes the empty list and the binary function symbol *cons* is the *list constructor*. In a term of the form $cons(x, y)$, x denotes the first element of the list and y denotes the rest of the list (which is itself a list). It is more convenient to use a more succinct representation for lists. PROLOG allows one to write

$$[a_1, a_2, \ldots, a_k]$$

instead of
$$cons(a_1, cons(a_2, \ldots cons(a_k, nil) \ldots)).$$
Furthermore,
$$[x|y]$$
is an abbreviation for
$$cons(x, y)$$
and [ ] stands for the empty list *nil*.

Example: $[[a, [b, c]] \mid [d, e]]$ is a shorthand for

$$cons(cons(a, cons(cons(b, cons(c, nil)), nil)), cons(d, cons(e, nil))).$$

The following diagram shows the structure of this term where each dot stands for an application of *cons*.

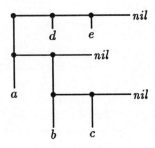

The most common operation on lists is the operation of concatenation (i.e. appending one list to another) which is denoted by *append*. The following logic program describes this operation.

$$append([\,], L, L).$$
$$append([X|L_1], L_2, [X|L_3]) \quad :- \quad append(L_1, L_2, L_3).$$

Here L, L_1, L_2, L_3 are variables, and $append(L_1, L_2, L_3)$ expresses the fact that L_3 is the concatenation of the lists L_1 and L_2.

Exercise 115: Present a successful computation of the above program with the given goal clause

$$?-\ append([a, b, c], [a, e, f], X).$$

What are the successful computations with the goal clause

$$?- \ append(X, Y, [a, b, c, a]).$$

Here X, Y are variables.

Exercise 116: Present a logic program that reverses a list. So, $reverse(L_1, L_2)$ should express that L_2 is the reversed version of L_1 (Example: ?– $reverse([1, [2, 3], 4, 5], Z)$ leads to the result $Z = [5, 4, [2, 3], 1])$.

Modify this program so that it reverses every sublist, too, not just the top level of the list. Call this predicate *deepreverse* (Example: ?– $deepreverse([1, [2, 3], 4, 5], Z)$ leads to the result $Z = [5, 4, [3, 2], 1])$.

Exercise 117: PROLOG has a built-in system predicate $atomic(X)$ that is evaluated successfully if and only if at the time of evaluation the variable X is instantiated with a constant. Formulate a logic program (using the predicates *atomic* and *add*) that computes the number of leaves of the binary tree that is represented by a list. For example,

$$?- \ leaves([a, [e, f], b, c], N).$$

should lead to the result $N = 5$.

Example: The following program is able to permute a list.

```
permute([ ], [ ]).
permute([X|Y], Z) :- permute(Y, W), insert(X, W, Z).
insert(A, B, [A|B]).
insert(A, [B|C], [B|D]) :- insert(A, C, D).
```

E.g., the goal clause

$$?- \ permute([it, never, rains, in, california], Z)$$

leads to the results

$$Z = [it, never, rains, in, california]$$
$$Z = [never, it, rains, in, california]$$
$$Z = [never, rains, it, in, california]$$
$$Z = [never, rains, in, it, california]$$

$$\vdots$$

Exercise 118: The following program is able to sort a list – but in a *very* inefficient way.

$$sort(L_1, L_2) \ :- \ permute(L_1, L_2), \ ord(L_2).$$
$$ord([\,]).$$
$$ord([X]).$$
$$ord([X|[Y|Z]]) \ :- \ X \leq Y, \ ord([Y|Z]).$$

Write a PROLOG program for Quicksort.

Example: The PROLOG program below translates arithmetical expressions, resp. assignments, into assembler code. For example, consider

$$z := (a * b) + c.$$

In this case we obtain

$$[[load, a], [load, b], mul, [load, c], add, [store, z]]$$

Here *load* means loading an element on a stack, and *mul* pops the top two stack elements, multiplies them, and pushes the result on top of the stack (*add* works analogously for addition). The command *store* stores the top stack element in the memory.

$$compile(X := Y, Z) \ :- \ compile(Y, W),$$
$$append(W, [[store, X]], Z).$$

$$compile(X * Y, Z) \ :- \ compile(X, U),$$
$$compile(Y, V),$$
$$append(U, V, W),$$
$$append(W, [mult], Z).$$

$$compile(X+Y,Z) \;:-\; compile(X,U),$$
$$compile(Y,V),$$
$$append(U,V,W),$$
$$append(W,[add],Z).$$

$$compile(X,[[load,X]]) \;:-\; atomic(X).$$

Exercise 119: First define formally the syntax of a programming language ASCA (a suitable subset of PASCAL), and then write a compiler for ASCA-programs in PROLOG.

Exercise 120: Implement a PROLOG interpreter in PASCAL that realizes PROLOG's depth-first search strategy, and additionally is able to handle at least some of PROLOG's built-in predicates.

PROLOG's incomplete depth-first evaluation strategy was already mentioned. Moreover, in Exercise 113 it was discussed that this incompleteness is, in a sense, inherent. It cannot be eliminated by rearranging the order of the clauses. Obviously this is a dilemma, and raises the question what possibilities there are to avoid such problems. PROLOG provides a somewhat peculiar way out of this dilemma, namely the *cut*.

Syntactically, the *cut* is like an atomic formula and is denoted by an exclamation mark (!). This type of atomic formula is only allowed on the right side of PROLOG procedure definitions (i.e. in the procedure body).

Example:
$$a \;:-\; b,c,!,d.$$

The presence of a *cut* does not influence the logic (i.e. semantics) of a clause, but the depth-first evaluation strategy is altered. Some part of the search tree will be cut off whenever such a *cut* symbol is present. By this, it is possible to ignore parts of the search tree that contain infinite computation paths (which are "dangerous" for the depth-first evaluation). On the other hand, the *cut* can just as well be "misused" to ignore parts of the search tree with existing solutions.

Here the opinions about the *cut* are split: Some programmers really want to have a tool in hand that allows influencing the search strategy and allows jumping over existing solutions (this technique will play a role below when we talk about the negation).

The other opinion is that the *cut* is in conflict with the ideal of logic programming. In logic programming languages the programmer should only specify *what* the problem to be solved is, but not *how* to solve it. A concept like the *cut* certainly belongs to the *how*-category. The *cut* in PROLOG is somehow comparable to the *goto* in standard operational programming languages.

How does the *cut* work in detail? Whenever a *cut* in a goal clause, like

$$?-!, a, b, c.$$

is evaluated for the first time, it leads to success immediately (as if there would be a fact consisting of "!." in the logic program). Then the next goal clause to evaluate would be

$$?- a, b, c.$$

But suppose the search process, by the backtracking mechanism, returns to the goal clause

$$?-!, a, b, c.$$

since there is no solution found in the subtree below the node $?- a, b, c.$ In this case the search process deviates from the depth-first search order. A "jump" is enforced that leads to the last parent goal clause in the search tree that did not contain the *cut*. This goal clause is then considered as evaluated non-successfully (that is, the Boolean variable *success* is set to **false** in the evaluation algorithm, cf. last section). This action has the consequence that potential subtrees located to the right of the node labeled with $?-!, a, b, c.$ are not considered – no matter whether they contain successful computations or not.

Example: Consider the logic program

$$a :- b, c.$$
$$b :- d, !, e.$$
$$d.$$
$$\vdots$$

and the goal clause ?− a. The following diagram demonstrates the whole search tree and the depth-first search order − and how the order is influenced by the *cut*.

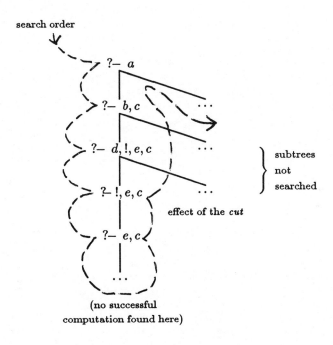

(no successful
computation found here)

Exercise 121: Modify the PROLOG evaluation algorithm from Section 3.2 such that the *cut* is correctly handled.

Exercise 122: The following piece of program

$$a \; :- \, b, !, c.$$
$$a \; :- \, d.$$

is used in actual PROLOG programs to simulate the well known **if-then-else** construct from standard programming languages. In a sense, the above program can be interpreted as

$$a \; :- \; \text{if } b \text{ then } c \text{ else } d.$$

Analyze the effect of the *cut* in the above program using SLD search trees.

Exercise 123: Consider the PROLOG program

happy	:–	*birthday, christmas.*
happy	:–	*birthday.*
happy.		
birthday	:–	*pigscanfly*
christmas	:–	*pigscanfly*
birthday	:–	*birthday*

(a) Construct the SLD search tree for the goal clause ?– *happy*.

(b) Rearrange the order of the clauses such that the depth-first search strategy finds a solution for the goal ?– *happy*.

(c) Insert a *cut* in the above program so that the (modified) search tree becomes finite – but as large as possible.

(d) Describe the effect of inserting a *cut* in any of the 3 possible positions in the first clause.

In the following we summarize the possible and typical applications of the *cut*.

1. After finding a first solution, an insertion of a *cut* allows one to forbid every further search for a solution. In some contexts it is clear that there is no other solution, or that the part of the tree not searched contains infinite computation paths. For example, in the logic program for addition, an insertion of a *cut* in the clause that expresses the base of the recursion will enforce that there is no further search whenever the first (and unique) solution has been found.

$$a(x, 0, x) \; :- \; !.$$
$$a(x, s(y), s(z)) \; :- \; a(x, y, z).$$

2. The *cut* allows construction clauses that perform a similar action as the **if-then-else** known from standard programming languages (cf. Exercise 122). The situation with the negation is similar (this is discussed in more detail below).

3. The sensible use of the *cut* allows one to improve the efficiency of programs because subtrees which are known to contain no solutions can be cut off in the search process. Although there is no general rule about where to introduce a *cut* to improve the efficiency – it depends very much on the intuition and experience of the programmer.

4. The *cut* allows one to overcome the logical incompleteness of PRO-LOG's depth-first evaluation strategy. Subtrees containing infinite computation paths can cut off.

Previously we discussed the notion that there are several reasons for restricting the predicate logic to Horn clauses in the PROLOG programming language. It was this restriction which permitted the procedural interpretation of logic programs, and which allowed us to develop such a finely structured theory (Clark's Completeness Theorem, the Swapping Lemma, and the various notions of semantics). But there can be cases where Horn formulas are too weak or not adequate to express the problem context. Here the *negation* plays a special role. Remember that the negation of a Horn formula in general is not equivalent to any Horn formula. But sometimes it is necessary to know whether a negative literal, say $\neg A$, follows from a logic program F. Viewed formally, this would correspond to a goal clause of the form ?– $\neg A$ or ?– $not(A)$. Surprisingly, the logical answer to such a question is always "no".

Exercise 124: Prove that there is no set of Horn clauses F and no negative literal $\neg A$ such that $\neg A$ is a consequence of F. In other words, Horn clause programs do not allow one to draw negative conclusions.

Therefore, our first attempt failed. But instead of asking the question whether the negation of A is a consequence of F, we now ask whether A is *not* a consequence of F. Obviously, in general this is not the same. In fact, to postulate equivalence of both notions means the same as claiming the *completeness* of the logic program F (in the sense discussed in Section 2.3), that is, for all closed formulas A, either $A \in Cons(F)$ or $\neg A \in Cons(F)$.

Assuming this condition holds, whenever A is not a consequence of F, then A is *false* (i.e. unsatisfiable). In this context, this condition is called the *closed world assumption*. Since A is *not* a consequence of F if and only if $F \wedge \neg A$ is satisfiable if and only if the empty clause is *not* derivable from the clause set $F \cup \{\{\neg A\}\}$, this situation is called *negation by failure*.

It is desirable to have at least this 'negation by failure' available in PROLOG. That is, if A cannot be proved to be a consequence of F, then assume that $\neg A$ is a consequence of F (and give the answer 'yes' on the query $?-\ \neg A$.) Still, this cannot be implemented – by basic principles: If for every A and F it could be determined in finite time whether A is a consequence of F, then the decidability of the predicate logic would follow. This contradicts the results obtained in Section 2.3. (The undecidability result still holds when we restrict ourselves to Horn clauses).

The next weaker form of negation is *negation by finite failure*. It means that $\neg A$ is a consequence of F is assumed if the SLD-computation tree of F with goal clause A is non-successful and *finite*. Exactly this form of negation is implemented in PROLOG: The goal clause $?-\ not(p)$ as input causes the PROLOG interpreter to search for successful computations of the form

$$(?-\ p, [\,]) \vdash_{F} \cdots \vdash_{F} (\square, sub).$$

Only if the search tree for this query is finite and does not contain a successful computation, the PROLOG interpreter outputs 'yes', otherwise 'no'. (Notice that apart from this, the interpreter cannot output any computation result). This form of negation is dangerous because the search for a successful computation might lead to an infinite path.

Negation by finite failure as introduced above can easily be expressed in PROLOG itself – 'misusing' the *cut* (existing solutions are cut off here).

$$not(P) \ :-\ P, !, fail.$$
$$not(P).$$

Here *fail* is a standard predicate whose evaluation always ends non-successfully so that backtracking occurs (just as if there is no program clause with procedure head *fail*). Observe that the variable P in the above program takes as values atomic formulas (instead elements of the Herbrand universe).

Exercise 125: Trace back the evaluation of the query $?-\ not(not(not(a)))$.

First suppose that a is a fact contained in the logic program, then that it is not.

Exercise 126: Consider the logic program

$$p(X) \; :- \, !, q(X).$$
$$p(a).$$
$$q(b).$$
$$q(a) \; :- q(a).$$

Find out what PROLOG answers to each of the following queries.

$$?- \; p(a).$$
$$?- \; not(p(a)).$$
$$?- \; q(a).$$
$$?- \; not(q(a)).$$
$$?- \; q(X), not(p(X)).$$
$$?- \; not(p(X)), q(X).$$

A further problem with PROLOG's form of negation is that in a goal clause of the form

$$?- \; \ldots not(t) \ldots$$

the term t (respectively the atomic formula t) should not contain any uninstantiated variables – at the time when t is evaluated. This can lead to an incorrect evaluation result. Consider for example the program

$$p(a).$$
$$q(b, b).$$

and the goal clause

$$?- \; not(p(X)), q(X, X).$$

The desired answer $X = b$ will not be obtained (Why?).

The situation is different with the goal clause

$$?-\ q(X, X), not(p(X)).$$

which leads to the answer $X = b$.

The intention of this section is to convince the reader of the fact that PROLOG is just *one* possibility for realizing the ideas of logic programming in the context of a usable programming language. It should be seen that other concepts are thinkable, and that the research in this direction is not yet settled.

Bibliography

T. Amble. *Logic Programming and Knowledge Engineering.* Addison-Wesley, Reading, MA, 1987.

K. R. Apt and M. H. van Emden. Contributions to the theory of logic programming. *Journal of the Association for Computing Machinery,* 29 (1977): 841–862.

G. Asser. *Einführung in die mathematische Logik I – III.* Verlag Harri Deutsch, Frankfurt/M, 1972.

M. Bauer, D. Brand, M. Fischer, A. Meyer, and M. Paterson. A note on disjunctive form tautologies. *SIGACT NEWS,* Vol. 5, No. 2 (1973): 17–20.

E. Bergmann and H. Noll. *Mathematische Logik mit Informatik-Anwendungen.* Springer-Verlag, Berlin, 1977.

W. Bibel. *Automated Theorem Proving.* Vieweg, Braunschweig, 1982.

W. Bibel and Ph. Jorrand (Eds.) *Fundamentals of Artificial Intelligence. Lecture Notes in Computer Science 232,* Springer Verlag, Berlin, 1985.

K.H. Bläsius and H.J. Bürckert (Eds.) *Deduktionssysteme.* Oldenburg Verlag, München 1987.

G.S. Boolos and R.C. Jeffrey. *Computability and Logic.* Cambridge University Press, Cambridge, 1974.

E. Börger. *Berechenbarkeit, Komplexität, Logik.* Vieweg, Braunschweig, 1985.

A. Bundy. *The Computer Modelling of Mathematical Reasoning.* Academic Press, London, 1983.

W.D. Burnham and A.R. Hall. *Prolog Programming and Applications.* Macmillan, London, 1985.

C. L. Chang and R. C. T. Lee. *Symbolic Logic and Mechanical Theorem Proving.* Academic Press, New York, 1973.

V. Chvátal and E. Szemerédi. Many hard examples for resolution. *Journal of the Assoc. Comput. Mach.* 35 (1988): 759–768.

K. L. Clark. *Predicate Logic as a Computational Formalism.* Research monograph 79/59 TOC, Imperial College, London, 1979.

K.L. Clark and S.-A. Tärnlund (Eds.) *Logic Programming,* Academic Press, New York, 1982.

W. F. Clocksin and C. S. Mellish. *Programming in Prolog.* Springer-Verlag, Berlin, 1981.

M.D. Davis and E.J. Weyuker. *Computability, Complexity and Languages,* Chapter 11+12. Academic Press, New York, 1983.

R.D. Dowsing, V.J. Rayward-Smith and C.D. Walter. *A First Course in Formal Logic and its Applications in Computer Science.* Blackwell Scientific Publ., Oxford, 1986.

B. Dreben and W. D. Goldfarb. *The Decision Problem – Solvable Classes of Quantificational Formulas.* Addison-Wesley, Reading, MA, 1979.

H. D. Ebbinghaus, J. Flum and W. Thomas. *Einführung in die mathematische Logik.* Wissenschaftliche Buchgesellschaft, Darmstadt, 1978.

M. H. van Emden and R.A. Kowalski. The semantics of predicate logic as a programming language. *Journal of the Association for Computing Machinery* 23 (1976): 733–742.

Y.L. Ershov and E.A. Palyutin. *Mathematical Logic.* Mir Publishers, Moscow, 1984.

D. M. Gabbey. *Elementary Logic – A Procedural Perspective.* Lecture notes, Imperial College, London, 1984.

J. H. Gallier. *Logic for Computer Science – Foundations of Automatic Theorem Proving.* Harper & Row, New York, 1986.

M. Gardner. *Logic Machines and Diagrams.* The University of Chicago Press, Chicago, 1958.

M. Genesreth, N. Nilsson. *Logical Foundations of Artificial Intelligence.* Morgan Kaufmann Publ., 1987.

F. Giannesini, H. Kanoui, R. Pasero and M. van Canegham. *PROLOG.* Addison-Wesley, Reading, MA, 1986.

C. Green. Theorem proving by resolution as a basis for question-answering systems. in B. Meltzer and D. Michie (Eds.) *Machine Intelligence 4*, 183–205, Elsevier Publ., New York, 1969.

M. Hanus. *Problemlösen in PROLOG.* Teubner, Stuttgart, 1986.

N. Heck. *Abstrakte Datentypen mit automatischen Implementierungen.* Dissertation, University of Kaiserslautern, 1984.

H. Hermes. *Einführung in die mathematische Logik.* Teubner Verlag, Stuttgart, 1976.

C.J. Hogger. *Introduction to Logic Programming.* Academic Press, New York, 1984.

J.E. Hopcroft and J.D. Ullman. *Introduction to Automata Theory, Languages, and Computation.* Addison-Wesley, Reading, MA, 1979.

A. Horn. On sentences which are true of direct unions of algebras. *Journ. of Symb. Logic* 16 (1951): 14–21.

O. Itzinger. *Methoden der Künstlichen Intelligenz,* Chapter 2. Carl Hanser Verlag, München, 1976.

M. Kaul. *Logik.* Lecture Notes, EWH Koblenz, 1983.

H. Kleine Büning and S. Schmitgen. *PROLOG.* Teubner Verlag, Stuttgart, 1986.

R. Kowalski. Predicate logic as programming language. *Information Processing* 74, 569–574, North-Holland, 1974.

R. Kowalski. *Logic for Problem Solving.* Elsevier North-Holland, Amsterdam, 1979.

R. Kowalski. Algorithm = Logic + Control. *Journal of the Association for Computing Machinery* 22 (1979): 424–436.

M. Levin. *Mathematical Logic for Computer Scientists.* Technical Report, MIT Project MAC, 1976.

H. R. Lewis and C. H. Papadimitriou. *Elements of the Theory of Computation*, Chapter 8–9. Prentice Hall, Englewood Cliffs, NJ, 1981.

H.R. Lewis. *Unsolvable Classes of Quantificational Formulas*. Addison-Wesley, Reading, MA, 1979.

J. W. Lloyd. *Foundations of Logic Programming*. Springer-Verlag, Berlin, 1984.

D. W. Loveland. *Automated Theorem Proving: A Logical Basis*. Elsevier North-Holland, New York, 1979.

Z. Manna. *Mathematical Theory of Computation*, Chapter 2. McGraw-Hill, New York, 1974.

J. Minker (Ed.) *Foundations of Deductive Databases and Logic Programming*. Morgan Kaufmann Publ., Los Altos, Ca., 1988.

L. Naish. *Negation and Control in PROLOG*. Lecture Notes in Computer Science 238, Springer, Berlin, 1986.

N.J. Nilsson. *Problem Solving Methods in Artificial Intelligence*, Chapter 6–8. McGraw-Hill, New York, 1971.

R. Nossum. Automated theorem proving methods. *BIT* 25 (1985): 51–64.

M.S. Paterson and M.N. Wegman. Linear Unification. *Journal of Computer and System Sciences*, 16 (1978): 158–167.

PROLOG. Special issue of *Communications of the Association for Computing Machinery*, 28, No. 12 (1985).

W. Rautenberg. *Nichtklassische Aussagenlogik*. Vieweg, Braunschweig, 1979.

M. M. Richter. *Logikkalküle*. Teubner Verlag, Stuttgart, 1978.

M. M. Richter. *Prinzipien der künstlichen Intelligenz*. Teubner Verlag, Stuttgart, 1989.

J. A. Robinson. *Logic: Form and Function*. Elsevier North-Holland, New York, 1979.

D. Rödding. *Einführung in die Prädikatenlogik*. Lecture Notes, University of Münster, 1970.

J. R. Shoenfield. *Mathematical Logic*. Addison Wesley, Reading, MA, 1967.

W. Schwabhäuser. *Modelltheorie* I + II. Bibl. Institut, Mannheim, 1971.

D. Siefkes. *Logik für Informatiker*. Lecture Notes, Techn. University of Berlin, 1986.

J. Siekmann and G. Wrightson (Eds.) *Automation of Reasoning 1 + 2*. Springer, Berlin, 1983.

L. Sterling and E. Shapiro. *The Art of Prolog*. MIT Press, Cambridge, Massachusetts, 1987.

A. Tarski, A. Mostowski and R. M. Robinson. *Undecidable Theories*. North-Holland, Amsterdam, 1971.

A. Thayse (Ed.) *From Standard Logic To Logic Programming*. Wiley, 1988.

R. Turner. *Logics for Artificial Intelligence*. Elis Horwood Limited, 1984.

A. Urquhart. Hard examples for resolution. *Journal of the Assocation of Computing Machinery* 34 (1987): 209–219.

T. Varga. *Mathematische Logik für Anfänger I + II*. Verlag Harri Deutsch, Frankfurt/M., 1972.

H. Vollmer. *Resolutionsverfeinerungen and ihre Vollständigkeitssätze*. study thesis, EWH Koblenz, 1987.

C. Walther. Automatisches Beweisen. in: *Künstliche Intelligenz*. Fachberichte Informatik 259, Springer, Berlin, 1987.

L. Wos. *Automated Reasoning – 33 Basic Research Problems*. Prentice-Hall, Englewood Cliffs, NJ, 1988.

L. Wos, R. Overbeek, E. Lusk, F. Boyle. *Automated Reasoning – Introduction and Applications*, Prentice-Hall, Englewood Cliffs, NJ, 1984.

Table of Notations

$\mathbb{N}$ is the set of natural numbers, including zero, $\mathbb{N} = \{0, 1, 2, \ldots\}$

$\{0, 1\}^*$ is the set of finite 0-1-strings, including the empty string
ε, $\{0, 1\}^* = \{\varepsilon, 0, 1, 00, 01, 10, 11, 000, \ldots\}$

$\{0, 1\}^+ = \{0, 1\}^* - \{\varepsilon\} = \{0, 1, 00, 01, 10, 11, 000, \ldots\}$

Notations defined in the text:

$\neg$	4	$Res^*(F)$	33, 89	$E(F)$	74
$\vee$	4	$\exists$	42	$[\,]$	84, 142
$\wedge$	14	$\forall$	42	$\forall H$	92
$\rightarrow$	4	F^*	42	$ANSWER$	109
$\leftrightarrow$	4	$Free(F)$	43	$:-$	115
$\bigwedge_{i=1}^{n}$	5	$U_\mathcal{A}$	44	$?-$	116
$\bigvee_{i=1}^{n}$	5	$I_\mathcal{A}$	44	$\vdash_F$	119
$\mathcal{A}$	5, 46	$f^\mathcal{A}$	44	S_{proc}	126
$\overline{\mathcal{A}}$	5	$P^\mathcal{A}$	44	S_{mod}	127
$\models$	9, 47	$x^\mathcal{A}$	44	Op_F	128
$\not\models$	9, 47	$\mathcal{A}_{[x/u]}$	46	Op_F^n	128
$\equiv$	14, 51	$[x/t]$	53	Fp_F	128
CNF	18	**RPF**	56	$S_{fixpoint}$	129
DNF	19	ε	64	$[x|y]$	142
$\square$	31	$Th(\mathcal{A})$	68	$!$	145
$Res(F)$	32, 89	$Cons(M)$	69	not	149, 150
$Res^n(F)$	33, 89	$D(F)$	70	$fail$	150

Index

Progress in Computer Science and Applied Logic

Progress in Computer Science and Applied Logic is a series that focuses on scientific work of interest to both logicians and computer scientists. Thus both applications of mathematical logic will be topics of interest. An additional area of interest is the foundations of computer science.

The series (previously known as *Progress in Computer Science)* publishes research monographs, graduate texts, polished lectures from seminars and lecture series, and proceedings of focused conferences in the above fields of interest. We encourage preparation of manuscripts in such forms as LaTeX or AMS TeX for delivery in camera-ready copy, which leads to rapid publication, or in electronic form for interfacing with laser printers or typesetters.

Proposals should be sent directly to the editors or to:
Birkhäuser Boston, 675 Massachusetts Ave., Suite 601, Cambridge, MA 02139

Progress in Computer Science and Applied Logic